Lecture Notes in Computer Science 8960

Commenced Publication in 1973
Founding and Former Series Editors:
Gerhard Goos, Juris Hartmanis, and Jan van Leeuwen

More information about this series at http://www.springer.com/series/8637

Abdelkader Hameurlain · Josef Küng
Roland Wagner · Tran Khanh Dang
Nam Thoai (Eds.)

Transactions on Large-Scale Data- and Knowledge-Centered Systems XVI

Selected Papers from ACOMP 2013

 Springer

Editors

Abdelkader Hameurlain
IRIT, Paul Sabatier University
Toulouse
France

Josef Küng
FAW, University of Linz
Linz
Austria

Roland Wagner
FAW, University of Linz
Linz
Austria

Tran Khanh Dang
Ho Chi Minh City University of Technology
Ho Chi Minh City
Vietnam

Nam Thoai
Ho Chi Minh City University of Technology
Ho Chi Minh City
Vietnam

ISSN 0302-9743   ISSN 1611-3349   (electronic)
Lecture Notes in Computer Science
ISBN 978-3-662-45946-1   ISBN 978-3-662-45947-8   (eBook)
DOI 10.1007/978-3-662-45947-8

Library of Congress Control Number: 2014958753

Springer Heidelberg New York Dordrecht London

Printed on acid-free paper

Springer-Verlag GmbH Berlin Heidelberg is part of Springer Science+Business Media
(www.springer.com)

Preface

The International Conference on Advanced Computing and Applications 2013 (ACOMP 2013) is an annual international forum for the exchange of ideas, techniques, and state-of-the-art applications in the field of advanced computing among scientists, engineers, and practitioners. For ACOMP 2013, we provided a friendly environment where researchers could discuss current and future trends in research areas such as Security and Information Systems, Software Engineering, Embedded Systems and VLSI Design, High Performance Computing, Image Processing and Visualization, Scientific Computing, and other interesting topics.

ACOMP 2013 received 96 submissions and, after a careful review process, only 30 papers were selected for presentation. Among those great papers, we selected seven papers to invite the authors to revise, extend, and resubmit for publication in this special issue. The main focus of this special issue is on advanced computing issues in information and security engineering as well as their promising applications.

The big success of ACOMP 2013, as well as this special issue of TLDKS, was the result of the efforts of many people, to whom we would like to express our gratitude. First, we would like to thank all authors who extended and submitted papers to this special issue. We would also like to thank the members of the committees and external reviewers for their timely reviewing and lively participation in the subsequent discussion in order to select such high-quality papers published in this issue. Finally yet importantly, we thank Gabriela Wagner for her enthusiastic help and support during the whole process of preparation for this publication.

October 2014

Tran Khanh Dang
Josef Küng
Nam Thoai

Organization

Editorial Board

Alberto Gambaruto	Barcelona Supercomputing Center, Spain
Eiji Kamioka	Shibaura Institute of Technology, Japan
Lam Son Le	University of Wollongong, Australia
Thanh Binh Nguyen	HCMC University of Technology, Vietnam
Yong-Meng Teo	National University of Singapore, Singapore
Minh-Quang Tran	National Institute of Informatics, Japan
Nguyen Huynh Tuong	HCMC University of Technology, Vietnam
Hoang Tam Vo	National University of Singapore, Singapore
Pham Tran Vu	HCMC University of Technology, Vietnam
Nguyen Ngoc Thien An	University College Dublin, Ireland
Viet-Hung Nguyen	University of Trento, Italy
Phan Trong Nhan	Johannes Kepler University Linz, Austria
Quoc Cuong To	Inria Rocquencourt, France
Le Thi Kim Tuyen	Sungkyunkwan University, South Korea

Contents

Visualizing Web Attack Scenarios in Space and Time Coordinate Systems. . . .   1
Tran Tri Dang and Tran Khanh Dang

Question-Answering for Agricultural Open Data.   15
Takahiro Kawamura and Akihiko Ohsuga

Learning-Oriented Question Recommendation Using Bloom's Learning
Taxonomy and Variable Length Hidden Markov Models.   29
Hilda Kosorus and Josef Küng

On the Performance of Triangulation-Based Multiple Shooting Method
for 2D Geometric Shortest Path Problems   45
Phan Thanh An, Nguyen Ngoc Hai, Tran Van Hoai, and Le Hong Trang

Protecting Biometric Features by Periodic Function-Based Transformation
and Fuzzy Vault.   57
Thu Thi Bao Le, Tran Khanh Dang, Quynh Chi Truong,
and Thi Ai Thao Nguyen

EPOBF: Energy Efficient Allocation of Virtual Machines in High
Performance Computing Cloud.   71
Nguyen Quang-Hung, Nam Thoai, and Nguyen Thanh Son

Human Object Classification Using Dual Tree Complex Wavelet
Transform and Zernike Moment.   87
Manish Khare, Nguyen Thanh Binh, and Rajneesh Kumar Srivastava

Erratum to: On the Performance of Triangulation-Based Multiple
Shooting Method for 2D Geometric Shortest Path Problems E1
Phan Thanh An, Nguyen Ngoc Hai, Tran Van Hoai, and Le Hong Trang

Author Index   103

Visualizing Web Attack Scenarios in Space and Time Coordinate Systems

Tran Tri Dang[✉] and Tran Khanh Dang

Ho Chi Minh City University of Technology, VNU-HCM, Ho Chi Minh City, Vietnam
{tridang,khanh}@cse.hcmut.edu.vn

Abstract. Intrusion Detection Systems can detect attacks and notify responsible people of these events automatically. However, seeing individual attacks, although useful, is often not enough to understand about the whole attacking process as well as the skills and motivations of the attackers. Attacking step is usually just a phase in the whole intrusion process, in which attackers gather information and prepare required conditions before executing it, and clear log records to hide their traces after executing it. Current approaches to constructing attack scenarios require pre-defining of cause and effect relationships between events, which is a difficult and time-consuming task. In this work, we exploit the linking nature between pages in web applications to propose an attack scenario construction technique without the need of cause and effect relationships pre-definition. Built scenarios are then visualized in space and time coordinate systems to support viewing and analysis. We also develop a prototype implementation based on the proposal and use it to experiment with different simulated attack scenarios.

Keywords: Web attack visualization · Attack scenario construction · Attack scenario visualization · Security visualization · Web security · Web attack understanding

1 Introduction

Although traditional Intrusion Detection Systems (IDSs) can detect individual attacks and notify responsible people of these events automatically, they usually lack the capability of synthesizing and presenting related attacks (or related events) to human users in an intuitive way. We argue that seeing separate attacks does not provide responsible people much support in understanding the big picture about the whole attacking process. This big picture contains not only the attack step, but also the information gathering and preparation steps before it, as well as the exploitation and identity hiding steps after it, among others. Therefore, having a tool to access this big picture is a real need for security administrators. Using this tool, the security administrators can see the whole intrusion process and may understand the motivations, the techniques, and the skills of attackers. This understanding is not only useful to make a counterattack immediately, but also to plan appropriate future defense strategies.

© Springer-Verlag Berlin Heidelberg 2014
A. Hameurlain et al. (Eds.): TLDKS XVI, LNCS 8960, pp. 1–14, 2014.
DOI: 10.1007/978-3-662-45947-8_1

To build the tool mentioned above, there are two problems that need to be solved. The first one is how to recognize events related to a particular attack to group them together. And the second one is how to display these events in an intuitive and comprehensible way. In this paper, we call the first problem "attack scenario construction", and the second problem "attack scenario visualization".

The attack scenario construction problem has been studied extensively in the past by network security researchers. Previous works proposed methods to construct attack scenarios using cause and effect relationships between events. That is, they define some conditions (called pre-conditions) that must be true for a particular attack to be valid, and they assume that when an attack is successful, it will make some other conditions (called consequences) to become true. The outputs of these works are attack scenarios that contain individual but related events in some orders. However, these approaches require that the administrators must define in advance all the cause and effect relationships between events, which is a difficult and time consuming task. Instead of doing so, in this paper, we focus on the security of web applications, and hence we are able to exploit the linking nature between web pages to propose an attack scenario construction technique without the need of cause and effect relationships pre-definition by a human user.

For the attack scenario visualization problem, although there are some studies about using information visualization techniques to present attacks, most of them focus on displaying these attacks separately. In other words, despite the attacks are displayed at the same time and on the same screen, usually no relationship information between the attacks (and other related events) is shown. And to the best of our knowledge, there is no published work about attack scenario visualization for web application domain yet. Thus, our hope is that the contributions of this work can provide some initial perspectives and ideas based on which further studies can be performed. To make our work more persuasive, we also develop a prototype implementation to demonstrate our proposal and use it to experiment with different attack scenarios. The experiment results show that our work can reveal some information that is not easy to see by using a traditional web application intrusion detection system.

To summarize, the contributions of this paper are as follow:

- Propose a technique to construct web-based attack scenarios by exploiting linking nature between web pages
- Propose an information visualization and user interaction technique to display attack scenarios to security administrators
- Implement a prototype for the above proposals and use it to experiment with attacks from both automatic tools and human attackers

The rest of the paper is structured as follow: in section 2, we review some related works; in section 3, we describe the architecture of our proposal; section 4 is about the visualization design to present attack scenarios; our experiments are reported in section 5; and section 6 concludes this paper with some plans for future works.

2   Related Works

Based on the level of required involvement of human security administrators, approaches to the detection of attacks on information systems can be classified into three categories: automatic, manual, and semi-automatic. Automatic solutions work without the need of constantly monitoring/controlling from human users and usually their outputs are somewhat simple, i.e. they raise alerts whenever they found something suspicious. It is the responsibility of the security administrators to investigate the alerts to verify if they are true, and to look for causes and effects of these alerts themselves. Some popular tools in this category include Snort [1] and Bro [2]. The main advantage of the automatic approach is that it requires the least human effort. However, because the output of this approach is simple, it does not help administrators to see the forest for the trees.

On the opposite side, techniques follow a manual approach rely on individual human administrators' own skills and knowledge for detecting, analyzing, and understand attacks. This approach is very limited and costs a significant human effort. Moreover, because the effectiveness of this approach depends on the individual administrators, it is not considered as a serious research topic.

At the middle point, semi-automatic solutions do some initial information processing, present the results (mostly in visual form) to people, get their interactions, and then repeat that information processing cycle. The important point here is that the human user is considered an essential component in this approach. This approach helps administrators to understand more clearly about security events that happen, because they have to interact with the systems to get needed results. Sometimes, the results from automatic IDSs can be utilized as a part of the input to these semi-automatic systems and are processed to provide a high level presentation of the overall security status, instead of as individual alerts [3] [4]. More specific works that target web application domain are [5] and [6].

To give administrators a bigger picture about their systems' security status, some works in network security field propose the use of correlation methods on intrusion alerts to combine them together in a meaningful manner. The first benefit of alerts correlation is that it reduces the examining effort of administrators by transforming a large set of individual alerts into a smaller set of related alerts, and allowing them to investigate these alerts at a higher level. The second benefit is that by grouping alerts together meaningfully, it provides a more comprehensible big picture to administrators. Usually, some relationships between alerts are specified in advance; and based on these relationships, correlation algorithms are used to group alerts together, as described in the work of Debar and Wespi [7]. In another work, Ning et al. use "prerequisites" and "consequences" to chain alerts together [8]. "Prerequisites" are conditions that must be satisfied for an attack to succeed, and "consequences" are possible outcomes once an attack is executed successfully. To show the correlation results in a more intuitive manner, the authors use visual directed graphs to display them. There is a common requirement of the correlation techniques used in [7] and [8] that is the rules used to combine alerts together must be defined in advance by the security administrators. Satisfying this requirement for medium to large information systems is a difficult and time-consuming task. Furthermore, when there are changes in the monitored system, these rules must be changed accordingly.

Clickstream analysis is another research area that is similar to our work of visually presenting users' activities on the web. The difference between our work and this research field lies in the main purposes of the two: our work focuses on security events, while the other focuses on normal users' activities. Currently, clickstream analysis is an important tool for electronic commerce websites owners. The merchants use it to know more about their website visitors, such as: where they come from, how they use the site, at which page they exit the site, at which page they decide to buy products, etc. Most of the time, clickstream is presented in visual forms and can interact with website owners. For example, Lee et al. propose two visualization methods for clickstream analysis: parallel coordinate and starfield graph [9]. Parallel coordinate is used to show the sequence of user steps happen on a website, such as look, click, buy, etc. It also shows the number of users drop ratios after each action step. Meanwhile, starfield graph is used to display product performance in terms of how many times they are viewed and how many times they are clicked after viewing. Another work by Kawamoto and Itoh try to integrate and visualize users' access patterns with existing website link structure for a better understanding about users' activities [10]. Commerce solutions in this area are also available, such as Google Analytics [11] and Webtrends [12].

3   System Architecture

Fig. 1 depicts the architecture of our web application attack scenario construction and visualization system. The system receives data inputs from web server access log, error log, IDS log, etc. This architecture is open so that new kinds of data can be added later to extend the system's capability. In addition to these data, which provide information about users' activities on web applications, the system also receives web applications' own structural data from an external crawler. This structural data contains information about linking relationships between web pages. Structural data and activities data are displayed in visual form to security administrators. Administrators can interact with the visual interface to get adjusted information. These interactions can affect both the scenario construction and scenario visualization components.

3.1   Data Collection and Preprocessing Component

This component collects users' activities data from various places. For each request that is recorded, it should contain data about access time, URI of accessed page, IP address, user agent, and query string. In our current prototype implementation, it just collects data from a web application IDS and an access log file generated by the Apache web server. In later versions, we intend to include more input sources. After collecting data, there is a preprocessing step that is used to ensure that all the data collected are standardized in a common format. Furthermore, because the collected data is spread over many places, this preprocessing component puts them in a central database for easy extracting and processing later.

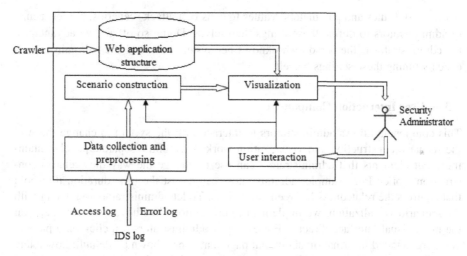

Fig. 1. The architecture of our proposed system. Inputs to the system include users' activities data stored in various log files, and web application structural data provided by an external crawler. Security administrators interact with the system through a visual interface. These interactions affect the working of both the scenario construction and scenario visualization components.

3.2 Scenario Construction Component

This component correlates information from various sources into attack scenarios. In this prototype, there are two input sources: HTTP requests to an Apache server, and alerts generated by a web application IDS. We assume that for a security administrator, she is more interested in alerts than normal requests. So, given a particular alert and some user input parameters, this component constructs two lists: a pre-events list and a post-events list. The pre-events list contains events (i.e. HTTP requests and IDS alerts) happen before the selected alert, and post-events list contains events happen after the selected alert. Dividing events into pre-events and post-events lists allows administrators to see more clearly about the preparation phase (containing pre-events data) and consequence phase (containing post-events data) for a particular attack. The construction of pre-events and post-events lists is similar to the session reconstruction task [13] found in the web usage mining research area. Both of them try to build a list of related events from a particular single event. But there is a major difference between our work and the session reconstruction task: the users' activities are assumed to be normal in session reconstruction task, while they are assumed to be anomalous in our work. From this assumption, it is clear that in our case, the users will try hard to hide their traces, making attack scenario construction more difficult than session reconstruction. Even in session reconstruction problem, it is reported that there are some difficulties that are not easy to solve [14]. Because of these difficulties, we do not use fixed rules and parameters' values to construct attack scenarios, but let security administrators to define these things themselves. Doing so offers two advantages: the administrator is the person who knows her system best, and she will gain experience by tuning these settings herself.

3.3   User Interaction Component

This component allows administrators to interact with the system to change the way the scenario construction and visualization work. We build a separate panel containing input elements that administrators can use to change the way the scenario construction works. For example, administrators can adjust the time duration threshold that controls the relatedness between two events. To let administrators to interact with the scenario visualization, we implement mouse actions like hover, selection, etc. on the main visual interface directly. For example, administrators can click on a page to get more detailed information about that page that is not shown by default: how many times it is accessed by users? How many alerts are generated on that page?

4   Visualization Design

The main purpose of the visualization component is to display the attack scenarios effectively to security administrators. Once an attack (i.e. an alert generated by a web application IDS) is selected, its related events are collected. For a true attack, usually there are other suspicious events happen together with it. But for a false attack, usually it may happen alone. By visualizing attacks and related events together, we think (but not have a proof yet) it is possible to detect false attacks from the real ones.

At the minimum, an event that is collected either from a web server log or from a web application IDS includes at least:

- The timestamp of the event
- The page URI that raises the event
- User agent information (IP address, agent name, version, etc.)
- Additional information (query string, response code, etc.)

In this work, the main goal of our visualization is to show how an attack scenario develops through time (when the chain of events happens) and space (where the chain of events accesses to). Seeing the attack scenarios in space and time coordinate systems can bring security administrators some information about what happen before an attack is launched (events in preparation phase), and what happen after that (events in consequence phase). In other words, administrators can learn some useful things about the attacking process an attacker uses: what does she do to prepare for an attack; what does she do if an attack is successful; what does she do to clear her traces, etc.

In the space-time visualization system that we use to visualize attack scenarios, there are two main visual areas: time-oriented coordinate systems area and space-oriented coordinate systems area. They are depicted in Fig. 2.

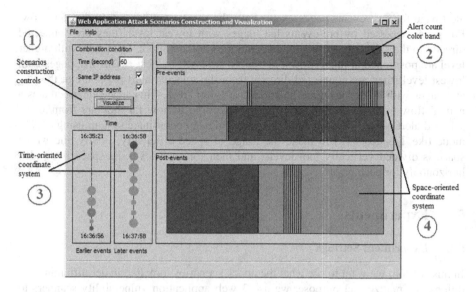

Fig. 2. Our system's visual interface design. There are four areas in it. Area #1 contains controls for interacting with the scenario construction. Area #2 contains the alert count color band for easy referencing by administrators. Area #3 contains time-oriented coordinate systems of events. And area #4 contains space-oriented coordinate systems of events.

4.1   Time-Oriented Coordinate System

In this coordinate system, related events of an attack are arranged in a one-dimensional time coordinate as depicted in the bottom left of Fig. 2. Each event is mapped to a circle whose center's location is determined by the event timestamp. To reduce clutter, events happen closed enough are grouped together into one bigger event. The size of each event circle depends on the number of elementary events (event that is recorded in a single entry of the web server log or in a single IDS record) contained in it. The color of each event circle is controlled by how much severity this event makes to the web application. In our implementation, this severity is calculated as the number of alerts generated in the event circle. This coordinate system helps security administrators to see how an attack scenario progresses through time.

4.2   Space-Oriented Coordinate System

In this coordinate system, events are grouped by the common URI they access. The purpose of this is to show how many events happen on each page and which pages receive the most interests from attackers compared to others. Similar to the time-oriented coordinate system, each page has its size determined by the number of events happen on it, and its color is determined by the severity degree it makes to the web application. We also group pages by their levels, which are the minimum number of links to reach to the selected alert page from that particular page for pre-events pages; or the minimum number of links to reach to that particular page from the selected alert page for post-events pages. Pages have the same level are put on the same row. For pages accessed by pre-events list, pages with lowest level (they are connected directly to the selected alert page) are positioned at the bottom, and pages with higher level are positioned at higher rows. For pages accessed by post-events list, pages with lowest level (they are connected directly from the selected alert page) are at the top, and pages with higher level are positioned at lower rows. This arrangement creates a natural flow from top to bottom of pages according to how far they are from/to the selected alert page. We implement the space-oriented coordinate system using a technique like Treemap [15] to make the usage of space efficiently. Firstly, the whole space is divided vertically into levels, and then for each level, its space is divided horizontally for each page.

5   Experiments

5.1   Experiment Settings

In this section, we want to see what the attack scenarios look like under different conditions. To realize that purpose, we use 3 web application vulnerability scanners to attack a web application that is installed on a local web server. We also invite a knowledgeable person in web application security to do some attacks on the same target. Our expectations for doing these tasks are that there are noticeable differences in the visualization results, and some knowledge about the attacking process can be reasoned from these visualizations.

The web server we use to host the test web application is an Apache HTTP server version 2.2.1 [16]. Accesses to this web server, either by the automatic scanning tools, or by the human attacker, are stored in an access.log file. We develop a script to extract the data from this log file and store them in a Java DB [17] database for later processing.

To capture attack/alert data, we use PHPIDS [18] as the web application IDS. Because PHPIDS can only detect intrusions to web applications written in PHP language, the target web application is a PHP-based one. The attacks that PHPIDS captures are stored in a MySQL database [19]. So we also create another script to copy them to the same Java DB database that stores Apache access records described previously. It is worth noting that PHPIDS, like other IDSs, definitely contains false detections, either negative or positive. As a result, we do not intend to evaluate our technique in term of correctness, but rather in term of insight it may bring to observers.

We also develop a custom crawler to extract linking relationships between pages on the target web application. These linking relationships are used to calculate page levels that are mentioned previously in the visualization design section. For two web pages A and B with two URIs UA and UB respectively, we say A links to B if either:

- The HTML content of A contains at least one hyperlink that points to U_B
- The HTML content of A contains at least one form that has its action attribute as U_B

After running the crawler, all linking relationships on the test web application are found and stored in the same Java DB that stores accesses and alerts data.

5.2   Experiment Scenarios

Using Automatic Tools

We use 3 web application vulnerability scanners, including Subgraph Vega [20], OWASP Zed Attack Proxy [21], and Acunetix Web Vulnerability Scanner [22], to attack the target web application. Among them, Subgraph Vega and OWASP Zed Attack Proxy are open source, while Acunetix Web Vulnerability Scanner is commercial, but it has a trial edition that can be used freely.

After using these tools to scan and attack the target web application, we randomly select an alert that is generated during the attacking process and construct a related attack scenario from it. Because we want to highlight both event sequences happen before and after the selected alert, the selected alert is chosen so that it is not among the ones generated too early or too late in the overall attacking process. We also set the attack scenario construction parameters to the same values for all these 3 cases to compare the 3 results more objectively. The visualization results of the attacks by Subgraph Vega, OWASP Zed Attack Proxy, and Acunetix Web Vulnerability Scanner are shown in Fig. 3, Fig. 4, and Fig. 5 respectively.

In Fig. 3, we can see that most of the attacks by Subgraph Vega happen around the selected alert, and these attacks in turn are preceded and followed by lots of other normal events, as shown in the time-oriented coordinate system. The attacks happen on a small number of web pages, as shown in the space-oriented coordinate system.

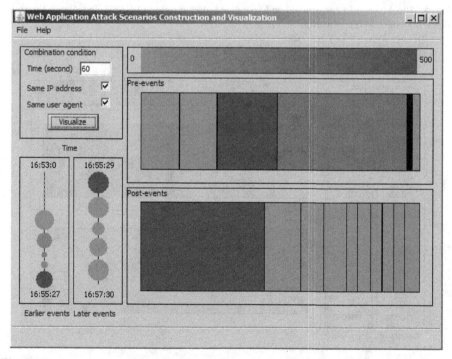

Fig. 3. Attack scenario constructed and visualized after running Subgraph Vega to attack the target web application

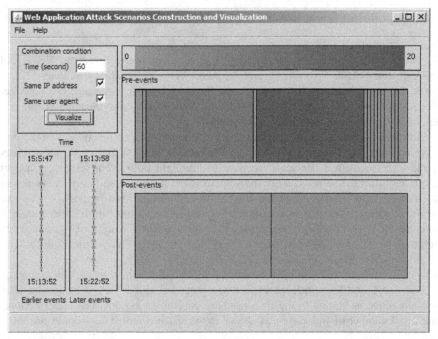

Fig. 4. Attack scenario constructed and visualized after running OWASP Zed Attack Proxy to attack the target web application

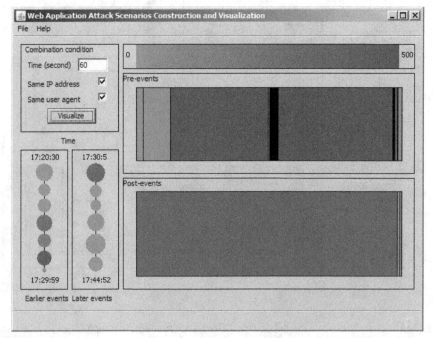

Fig. 5. Attack scenario constructed and visualized after running Acunetix to attack the target web application

In Fig. 4, we can see that OWASP Zed Attack Proxy attacks the target web application with a relatively uniform and slow speed. The number of normal accesses and recognized attacks in this case is small compared to the previous case. One feature similar to Subgraph Vega is that this tool also attacks a small number of pages.

Fig. 5 shows us that Acunetix adopts an attack strategy which is more aggressive than OWASP Zed Attack Proxy's, and somewhat similar to Subgraph Vega's, as depicted in the number of alerts generated. The Acunetix attack process looks rather regularly in time domain. Like the 2 previous tools, it also concentrates real attacks on a small number of web pages.

Manual Attack by a Human

We invite a human security expert to research and attack the target web application. We use different attack scenario construction parameters' values in this case because it is clear that the time period between requests will be longer when the requests are made by a human user. The visualization of the constructed attack scenario is displayed in Fig. 6. It is not surprised to see there is only a small number of normal accesses and attacks in Fig. 6. This result suggests that our technique is quite effective in distinguishing between attacks by automatic tools and attacks by human users. But we wonder about its capability in discriminating between attacks by an expert hacker and attacks by a script kiddie.

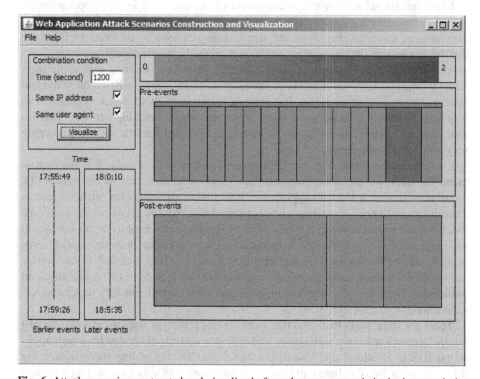

Fig. 6. Attack scenario constructed and visualized after a human person is invited to attack the target web application

6   Conclusions and Future Works

In this paper, we have proposed and implemented a web application scenarios construction and visualization technique to support security administrators in understanding intrusion processes happen on their systems. Unlike previous works on attack
scenarios construction, our work exploits the time constraints between related web
requests and the space constraints (linking relationships) between pages on web applications to build attack scenarios. As a result, it does not require cause and effect relationships manually defined in advance like other approaches. This feature helps security administrators reduce their effort significantly, especially in large web applications. On the visualization part, the main benefit our technique provides to security
administrators is that it can display not only individual attacks/alerts but also their
chains of events in a holistic way. In this way, events are positioned in time and space
coordinate systems based on when (time) and where (URL) they execute. Based on a
selected alert in the attack scenario under construction, we classify events into pre-
events (events happen before the event that raises the selected alert), and post-events
(events happen after the event that raises the selected alert). Our assumption in doing
so is that it is more helpful for security administrators to have a clear picture of what
constitute preparation steps (pre-events) and what constitute cleaning steps (post-
events).

Experiment results show the effectiveness of our approach in web application attack understanding for security administrators. By observing the visualization results,
we can have some insights about the attack strategies used by different tools/people
that are not easy to obtain by using traditional methods. In this regard, we believe our
proposed technique is a valuable complementary element to existing web application
IDSs: it offers security administrators meaningful attack scenarios based on individual
attacks detected by other IDSs. The lessons learnt from these attack scenarios can
help administrators to understand more about the whole attacking process. This information then can be used not only to immediately counter-attack but also to prepare
for future defense strategies.

There are some limits that are not addressed in this work. Firstly, it is the experiments in which we use some automatic tools and invite an expert to generate HTTP
requests and attack a test web application. All of them are somewhat artificial, and as
a result, are not very convincing when compared to a real life scenario. To overcome
this limit, we intend to setup a honey pot to attract real attackers from the Internet to
get more realistic data for later experiments. Secondly, our prototype lacks some
evaluations and feedbacks from real people that may use it in their day to day works,
i.e. security administrators. They can evaluate our work in term of its usability and
usefulness. There are other ways to measure our work though, but they require significant efforts in executing [23].

There are some ways to extend this work. In this prototype implementation, we just
use the access log from a web server and intrusion records generated by a web application IDS to construct and visualize attack scenarios. By adding more related log
data (e.g. error log, database calls log, etc.) and more IDS data (e.g. adding a second
IDS, etc.) for the visualization, our work may offer more useful information. This is

similar to the works of Erbacher et al. [3] and Livnat et al [4]. However, the works of Erbacher et al. and Livnat et al. are more about the current security status and do not provide the capability to view a whole attacking process as ours. The second way to extend this work is to integrate other information, such as alert type, severity, HTTP response code, etc. directly into the visualization space (currently they are displayed on demand via user interactions, i.e. mouse clicking). Doing so can save some analysis time of administrators and may offer more perspectives to them. The third way to extend this work is to use animation to replay attack scenarios (with appropriate time scaling methods to reduce/increase watching time). We believe using animation in this tool can provide more detailed information about attack scenarios. And last but not least, in our opinion, animation is more fun for administrators to work with.

Acknowledgement. This research is funded by Vietnam National University HoChiMinh City (VNU-HCM) under grant number C2013-20-08.

References

1. Roesch, M.: Snort-Lightweight Intrusion Detection For Networks. In: 13th USENIX Conference on System Administration, pp. 229–238. USENIX Association (1999)
2. Paxson, V.: Bro: a System for Detecting Network Intruders in Real-time. Computer Networks **31**, 2435–2463 (1999)
3. Erbacher, R.F., Walker, K.L., Frincke, D.A.: Intrusion and Misuse Detection in Large-Scale Systems. IEEE Computer Graphics and Applications **22**(1), 38–47 (2002)
4. Livnat, Y., Agutter, J., Moon, S., Erbacher, R.F., Foresti, S.: A Visualization Paradigm for Network Intrusion Detection. In: 6th Annual IEEE SMC Information Assurance Workshop, pp. 92–99. IEEE, West Point (2005)
5. Dang, T.T., Dang, T.K.: Visualization of Web Form Submissions for Security Analysis. International Journal of Web Information Systems Information **9**(2), 165–180 (2013)
6. Dang, T.T., Dang, T.K.: A Visual Model for Web Applications Security Monitoring. In: 2011 IEEE International Conference on Information Security and Intelligence Control, pp. 158–162. IEEE (2011)
7. Debar, H., Wespi, A.: Aggregation and Correlation of Intrusion-Detection Alerts. In: Lee, W., Mé, L., Wespi, A. (eds.) RAID 2001. LNCS, vol. 2212, pp. 85–103. Springer, Heidelberg (2001)
8. Ning, P., Cui, Y., Reeves, D.S.: Constructing Attack Scenarios Through Correlation of Intrusion Alerts. In: 9th ACM Conference on Computer and Communications Security, pp. 245–254. ACM, New York (2002)
9. Lee, J., Podlaseck, M., Schonberg, E., Hoch, R.: Visualization and Analysis of Clickstream Data of Online Stores for Understanding Web Merchandising. Data Mining and Knowledge Discovery **5**, 59–84 (2001)
10. Kawamoto, M., Itoh, T.: A Visualization Technique for Access Patterns and Link Structures of Web Sites. In: 14th International Conference Information Visualization, pp. 11–16. IEEE Computer Society, Washington (2010)
11. Google Analytics. http://www.google.com/analytics/
12. Webtrends. http://webtrends.com/

13. Spiliopoulou, M., Mobasher, B., Berendt, B., Nakagawa, M.: A Framework for the Evaluation of Session Reconstruction Heuristics in Web-Usage Analysis. INFORMS Journal on Computing **15**, 171–190 (2003)
14. Srivastava, J., Cooley, R., Deshpande, M., Tan, P.N.: Web Usage Mining: Discovery and Applications of Usage Patterns from Web Data. SIGKDD Explorations Newsletter **1**, 12–23 (2000)
15. Johnson, B., Shneiderman, B.: Tree-Maps: a Space-Filling Approach to the Visualization of Hierarchical Information Structures. In: 2nd Conference on Visualization, pp. 284–291. IEEE Computer Society Press, Los Alamitos (1991)
16. Apache HTTP Server. http://httpd.apache.org/
17. Java DB. http://www.oracle.com/technetwork/java/javadb/index.html
18. PHPIDS. http://www.phpids.org/
19. MySQL. http://www.mysql.com/
20. Subgraph Vega. http://subgraph.com/products.html
21. OWASP Zed Attack Proxy Project. https://www.owasp.org/index.php/OWASP_Zed_Attack_Proxy_Project
22. Acunetix Web Vulnerability Scanner. http://www.acunetix.com/
23. Plaisant, C.: The Challenge of Information Visualization Evaluation. In: International Working Conference on Advanced Visual Interfaces, pp. 109–116. ACM, New York (2004)

Question-Answering for Agricultural Open Data

Takahiro Kawamura[1,2](✉) and Akihiko Ohsuga[2]

[1] Corporate Research & Development Center,
Toshiba Corporation, Tokyo, Japan
takahiron.kawamura@toshiba.co.jp
[2] Graduate School of Information Systems,
University of Electro-Communications, Tokyo, Japan

Abstract. In the agricultural sector, the improvement of productivity and quality with respect to such attributes as safety, security and taste has been required in recent years. We aim to contribute to such improvement through the application of Information and Communication Technology (ICT). In this paper, we first propose a model of agricultural knowledge by Linked Open Data (LOD) with a view to establishing an open standard for agricultural data, allowing flexible schemas based on ontology alignment. We also present a semi-automatic mechanism that we developed to extract agricultural knowledge from the Web, which involves a bootstrapping method and dependency parsing, and confirmed a certain degree of accuracy. Moreover, we present a voice-controlled question-answering system that we developed for the LOD using triplification of query sentences and graph pattern matching of the triples. Finally, we confirm through a use case that users can obtain the necessary knowledge for several problems encountered in the agricultural workplace.

Keywords: Question-answering system · Linked Open Data · Agriculture

1 Introduction

Concern about food shortages affecting people in various parts of the world has been rising in recent years. However, since the expansion of the cultivated area is subject to constraints, it is necessary to improve productivity. On the other hand, the improvement of quality (safety, security, taste) is required in order to raise the incomes of farming households. In these circumstances, we aim to contribute to agricultural production by applying ICT techniques. Unlike producers involved in industrial production, most farmers, except for exemplary ones, are often confronted by unanticipated questions pertaining to several activities ranging from planting to harvesting, since agricultural work depends on complicated environmental factors. Thus, research[1] has been conducted with a view to developing a search engine to find previous problems similar to current problems. However, searching the Internet using a smartphone or a tablet PC on site

© Springer-Verlag Berlin Heidelberg 2014
A. Hameurlain et al. (Eds.): TLDKS XVI, LNCS 8960, pp. 15–28, 2014.
DOI: 10.1007/978-3-662-45947-8_2

is disadvantageous in that the user must input keywords and iteratively tap and scroll through a Search Engine Result Page (SERP) to find an answer. Therefore, this paper proposes a voice-controlled question-answering system for search of agricultural information. Voice control is suitable for agricultural work since users typically have dirty hands and can speak freely without disturbing other people. At the same time, it provides a mechanism for registering the work of the user, since data logging is the basis of precision farming according to the Japanese Ministry of Agriculture.

There are many sources of agricultural information on the Web. There are also many databases (DBs) authorized by agricultural organizations. However, the sources on the Web are written in natural language and the DBs do not employ uniform schema, making it impossible to search them using standardized procedures. Moreover, the DBs may have open application programming interfaces (APIs) for search, but contents are closed in many cases. Therefore, we propose Linked Open Data (LOD) for agricultural knowledge with a view to establishing an open standard for agricultural data.

This remainder of this paper is organized as follows. Section 2 introduces related work regarding standardization of agricultural data and applications supporting the work. In section 3, we propose Plant Cultivation LOD and describe a mechanism of LOD extraction from the Web. Then, in section 4, we describe the development and evaluation of the question-answering system for agricultural work using LOD. Finally, we conclude by referring to future work in section 5.

2   Related Work

A standard for agricultural data, agroXML[2], which is an XML schema for describing agricultural work, has been proposed. It is used as a means of exchanging data in a structured and standardized way between farm and external stakeholders (government, manufacturer, and retailer) in the EU. In practice, only the necessary part of the schema is exchanged according to the purpose. However, elements such as "cultivation" and "WorkProcess" in agroXML are mainly for data logging, and not for the description of cultivation knowledge. agroXML also adopts a hierarchical XML schema, making it difficult to trace content such as graphs (although the graph version is under development).

Regarding semantic application of agricultural data, the Food and Agriculture Organization (FAO) of the United Nations[3] is currently developing the Agricultural Ontology Service Concept Server, whose purpose is the conversion of the current AGROVOC thesaurus to Web Ontology Language (OWL) ontologies. AGROVOC is a vocabulary containing 40000 concepts in 22 languages covering agricultural subject fields, and expressed in the W3C Simple Knowledge Organization System (SKOS) and also published as LOD. To the best of our knowledge, however, AGROVOC does not include knowledge of plant cultivation.

With regard to filed applications for agriculture, Fujitsu Ltd. offers a recording system that allows the user to simply register work types by buttons on

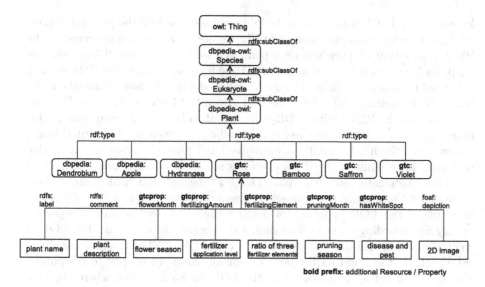

Fig. 1. Plant cultivation LOD

a screen with photos of the cultivated plants. NEC Corp. offers a machine-to-machine (M2M) service for visualizing sensor information and supporting farming diaries. Both systems address recording and visualization of the work, although our system is aimed at the search of cultivation knowledge on site by means of a voice-controlled question-answering system.

Recently, Apple's Siri has drawn attention to the question-answering system. Siri offers a high-accuracy voice recognition function and correctly answers the question in the case that the information source is a well-structured website such as Wikipedia. However, extracting the information from unstructured web sites often fails and Siri returns the search engine results page, and then the user needs to tap URLs from the list. Thus, LOD is a promising source of the question-answering system. In fact, IBM's Watson uses Linked Data as the internal information source in part [4,5]. [6] serves as a useful reference for surveying other question-answering systems, some of which uses LOD as the knowledge source. Although our proposed system is related to a number of works, it is distinguished by accuracy improvement and data acquisition by the user feedback and registration mechanism described in section 4. Also, there is no similar work in terms of agricultural application.

3   Open Data for Agricultural Information

3.1   Plant Cultivation LOD

Although the agroXML approach is ideal, adhering to a unique schema would pose difficulties for broader use. Therefore, we propose modeling cultivation knowledge by LOD, which allows flexible schemas based on the premise of their alignment using ontologies such as OWL. Figure 1 presents an overview of the Plant Cultivation LOD, where each plant is an instance of the "Plant" class of DBpedia[7] to which we refer as a base. DBpedia is a Linked Open Data dataset extracted from part of Wikipedia content. DBpedia is well known throughout the world. By September 2011, DBpedia had grown to 31 billion Resource Description Framework (RDF) triples. DBpedia has already defined more than 10,000 plants as types of the Plant class and its subclasses such as "FloweringPlant", "Moss" and "Fern". In addition, we defined 104 plants, mainly species native to Japan. Each plant of the Plant class has almost 300 Properties, but most of them are biologically inherited from "Thing", "Species" and "Eukaryote". Thus, we added 67 properties to represent necessary attributes for plant cultivation. In terms of the LOD Schemas for logging the work, we prepared Properties mainly for recording dates of flowering, fertilizing, and harvesting. The LOD is written in RDF, and currently stored in a cloud DB, DYDRA, whose SPARQL (SPARQL Protocol and RDF Query Language) endpoint is open to the public. The following listing is a snippet of the LOD in N3 notation, where the first column indicates a property name, and the second column is a brief description of the property.

```
@prefix dbpedia-owl: <http://dbpedia.org/ontology/>
@prefix dbpprop: <http://dbpedia.org/property/>
@prefix gtcprop: <http://www.uec.ac.jp/gtc/property/>

<http://dbpedia.org/resource/basil> a dbpedia-owl:Plant;
rdfs:subClassOf            depedia-owl:Eukaryote;
rdfs:label                 "Japanese name", "English name";
dbpprop:regionalOrigins    "Asia";
rdfs:comment               "Basil, or Sweet Basil, is a common name for
  ...";
foaf:page                  "reference page (url)";
foaf:depiction             "picture (url)";
gtcprop:priceValue         "market price";

# For cultivation knowledge
gtcprop:sunlight           "degree of illuminance";
gtcprop:perennial          'true' | 'false';
gtcprop:difficulty         "cultivation difficulty";
gtcprop:soil               "type of soil";

gtcprop:lowestTemperature  "MIN temperature for growth";
gtcprop:highestTemperature "MAX temperature for growth;
gtcprop:wateringAmount     "degree of watering";
gtcprop:plantingMonth      "start month for planting";
gtcprop:plantingMonthEnd   "end month for planting";
gtcprop:flowerMonth        "start month of blooming";
gtcprop:flowerMonthEnd     "end month of blooming";
gtcprop:fertilizingAmount  "degree of fertilization";
gtcprop:fertilizingMonth   "season of fertilization";
gtcprop:fertilizingElement "chemical elements of fertilization";
gtcprop:pruningMonth       "season for pruning";
gtcprop:pruningWay         "method of pruning";
gtcprop:fruitMonth         "season of harvesting";


# For disease and pest
gtcprop:hasWhiteSpot   "possible reason for the case (wikipedia uri)";
gtcprop:hasBlackSpot   "possible reason for the case (wikipedia uri)";
gtcprop:hasBrownSpot   "possible reason for the case (wikipedia uri)";
```

```
gtcprop:hasYellowSpot "possible reason for the case (wikipedia uri)";
gtcprop:hasMosaic     "possible reason for the case (wikipedia uri)";
gtcprop:hasFade       "possible reason for the case (wikipedia uri)";
gtcprop:hasKnot       "possible reason for the case (wikipedia uri)";
gtcprop:hasMold       "possible reason for the case (wikipedia uri)";
gtcprop:hasInsect     "possible reason for the case (wikipedia uri)";
gtcprop:hasNoFlower   "possible reason for the case (wikipedia uri)";

# For work logging
gtcprop:plantingSpace        "indoor or outdoor";
gtcprop:plantingDateTime     "date and hour of planting";
gtcprop:plantingAddress      "address";
gtcprop:plantingWeather      "weather";
gtcprop:plantingHighTemp     "highest temperature of the day";
gtcprop:plantingLowTemp      "lowest temperature of the day";

gtcprop:flowerSpace          "indoor or outdoor";
gtcprop:flowerDateTime       "date and hour of blooming";
gtcprop:flowerAddress        "address";
gtcprop:flowerWeather        "weather";
gtcprop:flowerHighTemp       "highest temperature of the day";
gtcprop:flowerLowTemp        "lowest temperature of the day";

gtcprop:wateringSpace        "indoor or outdoor";
gtcprop:wateringDateTime     "date and hour of watering";
gtcprop:wateringAddress      "address";
gtcprop:wateringWeather      "weather";
gtcprop:wateringHighTemp     "highest temperature of the day";
gtcprop:wateringLowTemp      "lowest temperature of the day";

gtcprop:fertilizingSpace     "indoor or outdoor";
gtcprop:fertilizingDateTime  "date and hour of fertilization";
gtcprop:fertilizingAddress   "address";
gtcprop:fertilizingWeather   "weather";
gtcprop:fertilizingHighTemp  "highest temperature of the day";
gtcprop:fertilizingLowTemp   "lowest temperature of the day";

gtcprop:purchaseSpace        "indoor or outdoor";
gtcprop:purchaseDateTime     "date and hour of purchase";
gtcprop:purchaseAddress      "address";
gtcprop:purchaseWeather      "weather";
gtcprop:purchaseHighTemp     "highest temperature of the day";
gtcprop:purchaseLowTemp      "lowest temperature of the day".
```

3.2 Agricultural Information Extraction from Web

This section describes a method for extracting cultivation knowledge from the Web, and constructing LOD. Our proposed method is inspired by [8] at AAAI10, which proposed a semi-automatic extraction service from the Web using the existing ontologies, where several learning methods are combined to reduce extraction errors. Although [8] focused on the world knowledge, and thus the granularity and the number of properties for each instance are rather abstract and limited, our method retains the variety of the properties and keeps the extraction accuracy by restricting the domain of interest.

LOD Extraction Method. We developed a semi-automatic method for growing the existing LOD to collect the necessary plant information from the Web and correlate it to DBpedia, which includes a dependency parsing method based on WOM Scouter[9] and a bootstrapping method based on ONTOMO[10]. In the plant information, the plant names are easily collected from a list on any gardening web sites, and also we have already defined the property names from the aspect of the plant cultivation. We thus need the value of the property for each plant. The process of our LOD extraction is shown in Fig. 2.

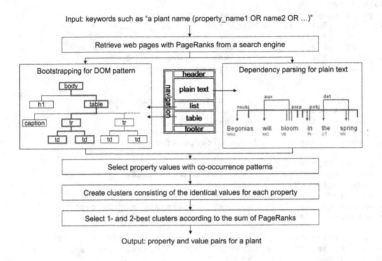

Fig. 2. Process of LOD content generation

We first create a keyword list, which consists of an instance name, that is, plant name and a logical disjunction of the property names, such as "basil" ("Japanese name" OR "English name" OR "country of origin" OR ...), and then search on a web search engine, and receive more than 100 web pages. We then retrieve the page contents, except for PDF files and also take a Google PageRank value for each page.

The bootstrapping method extracts specific patterns of the document object model (DOM) tree in the page contents using some keys, which are the property names or their synonyms, and then applies the patterns to other web pages for the extraction of other property values. The method is used for the extraction from structured parts of the page contents like tables and lists.

There, however, are a number of gardening web sites, where most of the page contents are described in plain text. We thus developed an extraction method using dependency parsing, since a triple $< plantname, property, value >$ is regarded as a dependency relation $< subject, verb, object >$. The method follows dependency relations in a sentence from a seed term, which is a plant name, a property name, or their synonym, and then extracts a triple, or a triple without a subject in the case of no subject within a sentence (the subject will be filled with a plant name in the keyword list later).

Next, we select property values that match with co-occurrence strings which are prepared for each property name, for example, the "temperature" property must match with °C or °F. We then create clusters of the identical property values for each property based on Longest Common Substring (LCS), add up the PageRank values of the source web pages in each cluster, in order to excludes errors of the extraction and of the information source, then to determine the best possible property value and the second-best. Experienced gardeners finally select a correct value for each property from the extracted values. If there are various theories as to the correct value for the property, they selected the dominant one.

LOD Extraction Accuracy. The LOD extraction method was evaluated for 13 properties values of 90 plants. Table 1 shows precisions and recalls (avg.) of the best possible value (1-best) separated by the whole process, the bootstrapping method, and the dependency parsing. The precisions and recalls of the second-best possible value (2-best) of the whole process is also shown in the table. Although we retrieved more than 100 web pages for each plant, DOM parse errors and difference of file types reduced the page amount to about 60%. In the case that the sum of the PageRank values of two clusters are the same, two values are regarded as the first position. In addition, the accuracy is calculated in units of the cluster instead of each extracted value. In the case of 1-best, a cluster which has the biggest PageRank value is an answer for the property. In the case of 2-best, the two biggest clusters are compared with a correct value, and if either of the answers is correct, it is regarded as correct. N-best precision is defined as follows:

$$N - best \ precision = \frac{1}{|D_q|} \sum_{1 \le k \le N} r_k$$

,where $|D_q|$ is the number of correct answers for a query q, and r_k is a function equaling 1 if the item at rank k is correct, zero otherwise.

The result of 1-best achieved a precision of 85% and a recall of 77%, and the 2-best achieved a precision of 97% and a recall of 87%. We thus confirmed that it is possible to present the binary choice including a correct answer in many cases. The automatic extraction will not be perfect after all, and then manual checking is necessary at the final step. Therefore, the binary choice is a realistic design. In more detail, the bootstrapping collected smaller amounts of values, and the recall was lower than the dependency parsing. However, the precision was higher than the dependency parsing. The reason is that data written in tables was correctly extracted, but lacks diversity of properties. The dependency parsing collected a large amount of values including many noisy data, and then the total accuracy was affected by the dependency parsing. The reason is that the biggest cluster of the PageRank value was composed of the values extracted by the dependency parsing. We thus plan to set some weights on the values extracted by the bootstrapping.

Table 1. Extraction accuracy (%)

Accuracy	1-best	2-best	1-best by bootstraping	1-best by dependency parsing
Precision	85.2	97.4	88.6	85.2
Recall	76.9	87.2	46.2	76.9
Amount Ratio	–	–	10.8	89.2

4   Question-Answering System for Agriculture

4.1   Problem and Approach

The basic operation of our question-answering system is extraction of a triple such as $subject, verb$, and $object$ from a query sentence by using morphological analysis and dependency parsing. Any question words (what, where, when, why, etc. are then replaced with a variable and the LOD DB is searched. In other words, the $< subject, verb, object >$ triples in the LOD DB are matched against $<?, verb, object >$, $< subject, ?, object >$, and $< subject, verb, ? >$ in the query. SPARQL is based on graph pattern matching, and this method corresponds to a basic graph pattern (one triple matching). At the data registration, if there is a resource corresponding to the $subject$ and a property corresponding to the $verb$ from the user statement, a triple that has the $object$ from the user statement as the value is added to the DB.

However, since the schema is open, mapping of query sentence to the schema poses a problem. Although mapping between the verb in the query sentence (in Japanese) and a Property in the LOD schema of the DB must be defined in advance, both of them are unknown in this open schema scenario (in the closed DB the schema is given), so the score according to the mapping degree cannot be predefined. The open schema means that the schema is not regulated by any organization, and there may be several properties of the same meaning and a sudden addition of a new property. In addition, we assume searching over multiple LOD sets made by the different authors. We therefore use a string similarity and a semantic similarity technique using the WordNet thesaurus from the field of ontology alignment to map verbs to Properties, and attempt to improve the mapping based on user feedback. We first register a certain set of mappings {verb, property} as seeds in the Key-Value Store (KVS). If a verb is unregistered, we then do the following:

(1) Expand the $verb$ to its synonyms using Japanese WordNet ontologies, and then calculate the LCS with the registered $verbs$ to use as the similarity.
(2) Translate the new $verb$ into English, and calculate the LCS of the English with the registered properties.
(3) If we find a resource that corresponds to a $subject$ in the query sentence in the LOD, we then calculate the LCSs of the translated $verbs$ with all the properties belonging to the resource, and create a ranking of possible mappings according to the combination of the above LCS values.

(4) The user feedback that indicates which property was actually viewed is sent to the server, and the corresponding mapping of the new *verb* to the property is registered in the KVS.

(5) Since the registered mappings are not necessarily correct, we recalculate the confidence value of the mapping based on the number of pieces of feedback, and update the ranking of the mapping to improve the N-best accuracy (see section 4.3).

Also, we provide a registration mechanism of the user context information to increase contents of the Plant Cultivation LOD. When the user registers a sentence in the DB, the sensor data are automatically aggregated by using built-in sensors on the smartphone, and the context information at that time and location is inserted into the DB. For example, when a user registers a triple describing a flower has blossomed, the sensor data for the location is converted to literal, one for the temperature is converted to integer, and one for the space is translated to Indoor or Outdoor, respectively. Then, the context information such as gtcprop:flowerAddress (location), gtcprop:flowerDateHighTemp (highest temperature of the day), gtcprop:flowerDateLowTemp (lowest temperature of the day), and gtcprop:flowerSpace (space of the flower) is automatically registered in the LOD DB. We prepared the LOD schemas (properties) corresponding to the context information (see 'work logging' part of the previous listing). Therefore, the user can register not only the direct assertion, but also several background information at once.

4.2   Development of Question-Answering System

Figure 3 shows an interface of our question-answering system. It automatically classifies the speech intention (Question Type) of the user into the following four types (Answer Type is a literal, URI, or image).

1. Information Search
   Search for plant information in the LOD DB.
2. Information Registration
   Register new information for a plant that does not currently exist in the LOD DB or add information to an existing plant.
3. Record Registration
   Register and share records of the daily work. However, the verbs that can be registered are limited to the predefined Properties in the LOD.
4. Record Search
   Search through records to review previous work and view the work of other people.

Figure 4 shows the architecture of our question-answering system. The user can input a query sentence by Google voice recognition or keyboard. The system then accesses the Yahoo! API for Japanese morphological analysis, extracts a triple using the built-in dependency parser, and generates a SPARQL query by filling in slots in a query template. The search results are received in XML format.

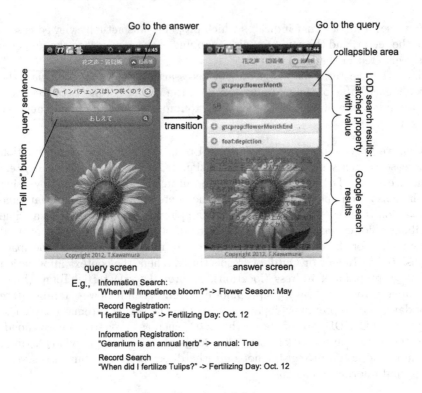

Fig. 3. Interface of QA system

After searching the {verb, property} mappings registered in Google Big Table and accessing the Microsoft Translator API and Japanese WordNet Ontology provided by the National Institute of Information and Communications Technology (NICT), the LCS values for each mapping are calculated as described in section 4.1. The order of matching is firstly matching the Subject against Resources by tracing 'sameAs' and 'wikiPageRedirects' links, and then searching for Verb matches with the Properties of the Resources. A list of possible answers is then created from the pairs of Properties and Values with the highest LCS values. The number of answers in the list is set to three owing to constraints on the client UI. The results of a Google search are also shown below in the client to clarify the advantages and limitations of the QA system by comparison. The user feedback is obtained by opening and closing a collapsible area in the client that gives a detailed look at the Value of the Property. During searches, the feedback updates the confidence value of a registered mapping {verb, property} or registers a new mapping. During registration, the feedback has the role of indicating which of three properties the object (value) should be registered to. The client UI displays the results. Text-to-speech has not been implemented yet. The query currently matches the graph pattern $< subject, verb, ? >$ only. As the target LOD, the QA system can search not only on the Plant Cultivation LOD, but also on DBpedia.

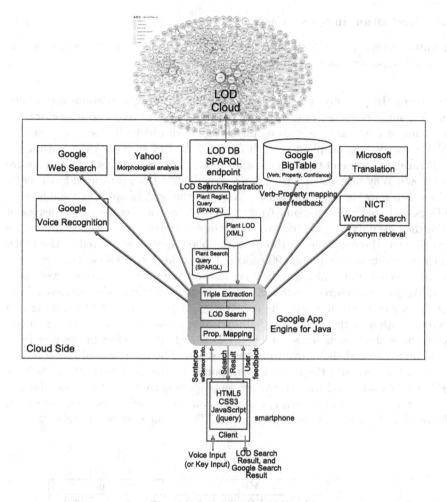

Fig. 4. Mashup architecture

The context registration mechanism is realized by the acquisition of sensor data. The sensor data are obtained by Android OS 2.2+ functions and jQuery 1.6.4+ running on the smartphone, and the related web services. For example, date and time information are obtained from the internal clock, and location is obtained by the Global Positioning System (GPS) function in the Android OS. But Point of Interest (POI) and weather information (temperature and humidity) are obtained by accessing Yahoo! Open Local Platform (olp.yahoo.co.jp), and Japan Meteorological Agency (www.jma.go.jp) based on the time and the GPS information. The POI specifies location names (buildings, companies, stations) around that location. Furthermore, we determine the location as indoor or outdoor space by the built-in illuminance sensor.

Our system is available at www.ohsuga.is.uec.ac.jp/~kawamura/fv.html (in Japanese) and won a Judges's Special Award in the LOD Challenge Japan 2012.

4.3   Evaluation in a Use Case

In this section, we set a specific use case and conduct an evaluation on the accuracy improvement and the context registration.

Accuracy Improvement. As a use case where the question-answering system is useful, we focused on a situation where the user asks the reason for the problems of plant growth such as physiological disorder, disease and pest. We collected questions on plant growth from Q&A sites such as Yahoo! Answers, and translated some of them to the spoken language as a dataset. Examples are as follows: Why have the leaves of black pine died? / Does corn need fertilizer? / Where is a suitable space for sage? / Why has the bark of apple become brown? / Does leaf curl affect passion fruit? / Why have some edges of the leaves of kalanchoe been dying? / Is tomato with white patterns safe to eat? However, we have limited the questions to those for the plant species registered in the LOD. In the experiment, we divided 90 collected questions into 10 sets. Then, we randomly selected and evaluated the first set and the next set consecutively as a test. We gave the correct feedback (the correct answer was also retrieved from the Q&A sites), which means the registration of {verb, property} mapping and incrementation of the confidence value, to one of the three answers per query. After the evaluation of the second set, we cleared all the effect of the user feedback, and repeated the above from the first set. The difference of the accuracy between the first and the second corresponds to the improvement of the user feedback. We assumed that the query sentence is correctly entered and did not consider voice recognition error, since we can select the correct sentence from the results of the Google voice recognition. The result is shown in table 2.

Table 2. Accuracy of search

	False		True	
	no Prop.	triplification error	1-best	3-best
1st Set (avg.)	22.2%	11.1%	55.6%	66.7%
2nd Set (avg.)			66.7%	66.7%

In the table, the result of False shows the average of the first set and the second set. We found that the coverage of the prepared Properties remains approx. 80% of the questions. We plan to expand the Properties defined in the Plant Cultivation LOD. In addition, the accuracy of the conversion from a sentence to a triple (triplification) was rather high, almost 90%. The current extraction mechanism is rule-based, but we intend to extend the rules and use of machine learning techniques to manage the broader questions. On the other hand, the result of True was about 67% of the accuracy in 3-best. By comparing the 1-best result for the first set with the second one, we can confirm that the problem raised in section 4.1, the mapping of the verb in the question to the LOD schema, has been improved about 10% by the user feedback. (Note that 1-best accuracy equals 3-best accuracy means all the correct answers are in the first position, that is, they are amongst the first three positions.) However, this use case only focuses on the questions related to ill-growth of the plant. Moreover, we evaluated the search function in this experiment, but the accuracy of the registration has not been studied yet. In the near future, we intend to collect a broader range of questions and registration statements, and conduct the evaluation on the accuracy and the performance.

Context Registration. We also conducted experiments to confirm the usefulness of the context registration. In the experiment, we also used the dataset regarding questions and answers on plant growth retrieved from the Q&A sites. Table 3 presents the properties representing a series of planting → flowering → pruning → withering that are registered by the user statement, and the context information obtained by the sensors. In the table, the intersection of the line of 'gtcprop:flowerDate' and the column of 'Location' means that when a user registers the date a flower has blossomed, then the Location information is automatically registered at the same time. Then, the number where the line and the column intersect means the co-occurrence ratio of these two information in the answers of the dataset. That is, answers for questions about the time when a flower blossoms include the Location information with 54.5% probability in the dataset. The size of the dataset is as follows: 414 answers for 'planting-Date', 114 answers for 'flowerDate', 128 answers for 'pruningDate', and 99 answers for 'dieDate'. But we excluded short answers consisting of only one or two lines. In this experiment, we considered that the additional information that has been described together with the original information in the answer is worth registering to the Plant Cultivation LOD. Therefore, we regarded that the automatic registration of such context information has usefulness.

Table 3. Co-occurrence ratio (%) of registered property and collected context

registered property	Time	Space {Indoor, Outdoor}	Location	Weather	High Temp.	Low Temp.	Humid
gtcprop:plantingDate	0.0	42.9	14.3	28.6	42.9	57.1	42.9
gtcprop:flowerDate	9.1	45.5	54.5	18.2	36.4	45.5	27.3
gtcprop:pruningDate	0.0	11.1	11.1	0.0	0.0	11.1	11.1
gtcprop:dieDate	0.0	12.5	25.0	25.0	25.0	37.5	37.5
(avg.)	2.3	28.0	26.2	17.9	26.1	37.8	29.7

As a result, the context information has the variation in its usefulness (0.0% – 57.0%). However, most of them have at least more than 20% on the average, thus the context registration mechanism can be regarded useful.

5   Conclusion and Future Work

This paper proposed the Plant Cultivation LOD with a view to establishing an open standard for agricultural data that models cultivation knowledge with flexible schemas without adhering to a hierarchical structure, and then presented a mechanism that we developed to extract the necessary knowledge from the Web and an evaluation of its accuracy. Moreover, we proposed a voice-controlled question-answering mechanism for this open schema LOD in order to obtain knowledge of the problems that farmers encounter on site, and then evaluated its accuracy through a use case. We are now planning to expand the Plant Cultivation LOD, and also considering conducting an evaluation of our question-answering system in the agricultural workplace.

References

1. Ninomiya, S.: ICT-enabled Agricultural Science for Development Scenarios, Opportunities, Issues. In: Proc. of Science Forum 2009, pp. 1-5 (2009)
2. Schmitz, M., Martini, D., Kunisch, M., Mosinger, H.J.: agroXML - Enabling Standardized, Platform-Independent Internet Data Exchange in Farm Management Information Systems. In: Metadata and Semantics, pp. 463–468 (2009)
3. Agricultural Information Management Standards. http://aims.fao.org/ (accessed: September 25, 2014)
4. Welty, C., Barker, K., Aroyo, L., Arora, S.: Query Driven Hypothesis Generation for Answering Queries over NLP Graphs. In: Cudré-Mauroux, P., Heflin, J., Sirin, E., Tudorache, T., Euzenat, J., Hauswirth, M., Parreira, J.X., Hendler, J., Schreiber, G., Bernstein, A., Blomqvist, E. (eds.) ISWC 2012, Part II. LNCS, vol. 7650, pp. 228–242. Springer, Heidelberg (2012)
5. Welty, C., Murdock, J.W., Kalyanpur, A., Fan, J.: A Comparison of Hard Filters and Soft Evidence for Answer Typing in Watson. In: Cudré-Mauroux, P., Heflin, J., Sirin, E., Tudorache, T., Euzenat, J., Hauswirth, M., Parreira, J.X., Hendler, J., Schreiber, G., Bernstein, A., Blomqvist, E. (eds.) ISWC 2012, Part II. LNCS, vol. 7650, pp. 243–256. Springer, Heidelberg (2012)
6. Lopez, V., Uren, V., Sabou, M., Motta, E.: Is question answering fit for the semantic web? A survey Semantic Web J. 2(2), 125–155 (2011)
7. Auer, S., Bizer, C., Kobilarov, G., Lehmann, J., Cyganiak, R., Ives, Z.G.: DBpedia: A Nucleus for a Web of Open Data. In: Aberer, K., Choi, K.-S., Noy, N., Allemang, D., Lee, K.-I., Nixon, L.J.B., Golbeck, J., Mika, P., Maynard, D., Mizoguchi, R., Schreiber, G., Cudré-Mauroux, P. (eds.) ASWC 2007 and ISWC 2007. LNCS, vol. 4825, pp. 722–735. Springer, Heidelberg (2007)
8. Carlson, A., Betteridge, J., Kisiel, B., Settles, B., Hruschka, Jr., E.R., Mitchell, T.M.: Toward an Architecture for Never-Ending Language Learning. In: Proc. of 25th Conference on Artificial Intelligence (AAAI), pp. 1306–1313 (2010)
9. Kawamura, T., Nagano, S., Inaba, M., Mizoguchi, Y.: Mobile Service for Reputation Extraction from Weblogs - Public Experiment and Evaluation. In: Proc. of 22nd Conference on Artificial Intelligence (AAAI), pp. 1365–1370 (2007)
10. Kawamura, T., Shin, I., Nakagawa, H., Nakayama, K., Tahara, Y., Ohsuga, A.: ONTOMO: Web Service for Ontology Building - Evaluation of Ontology Recommendation using Named Entity Extraction. In: Proc. of International Conference WWW/INTERNET 2010 (ICWI), pp. 101–111 (2010)

Learning-Oriented Question Recommendation Using Bloom's Learning Taxonomy and Variable Length Hidden Markov Models

Hilda Kosorus$^{(\boxtimes)}$ and Josef Küng

Institute of Application Oriented Knowledge Processing,
Johannes Kepler University, Linz, Austria
{hkosorus,jkueng}@faw.jku.at
http://www.faw.jku.at

Abstract. The information overload in the past two decades has enabled question-answering (QA) systems to accumulate large amounts of textual fragments that reflect human knowledge. Therefore, such systems have become not just a source for information retrieval, but also a means towards a unique learning experience. Recently developed recommendation techniques for search engine queries try to leverage the order in which users navigate through them. Although a similar approach might improve the learning experience with QA systems, questions would still be considered as abstract objects, without any content or meaning. In this paper, a new learning-oriented technique is defined that exploits not only the user's history log, but also two important question attributes that reflect its content and purpose: the topic and the learning objective. In order to do this, a domain-specific topic-taxonomy and Bloom's learning framework is employed, whereas for modeling the order in which questions are selected, variable length Markov chains (VLMC) are used. Results show that the learning-oriented recommender can provide more useful, meaningful recommendations for a better learning experience than other predictive models.

Keywords: Recommender systems · Question-answering systems · Learning taxonomy · Topic taxonomy · Collaborative filtering · Variable length markov chain

1 Introduction

In the past two decades, education has become more and more subject to personalization and automation. The success of recommender systems [1] has motivated research on deploying such techniques also in educational environments to facilitate access to a wide spectrum of information [16].

One of the consequences of information overload is the rise of question answering (QA) systems. Over time, QA systems have gathered a large amount of textual fragments – reflections of human knowledge - from a variety of domains

© Springer-Verlag Berlin Heidelberg 2014
A. Hameurlain et al. (Eds.): TLDKS XVI, LNCS 8960, pp. 29–44, 2014.
DOI: 10.1007/978-3-662-45947-8_3

and, therefore, represent a potential source for learning and establishment of new fields of study. Such systems have become not just a source of information retrieval, but also a medium for online information seeking and knowledge sharing [15], a means towards a unique learning experience. However, the exponential growth in the data volume of QA systems has made the users access to the desired information more difficult and time-consuming [15].

Current QA systems integrate traditional content-based recommendation engines with the goal to identify the most suitable user to answer a question [11,12], but little research aims at filtering out for the user the questions/answers that might be of interest [15]. The drawback of such approaches is that the recommender does not take into account explicitly the user's learning goals or learning process, neither the order in which questions are selected. The goal of this paper is to improve the learning experience of the user in the role of question asker.

Recent research in the field of query recommendation for search engines are based on query search graphs that aim at extracting interesting relations from user query logs [3,4]. Some of these graphs are constructed based on relations between queries, which are explored and categorized according to different sources of information (e.g., words in a query, clicked URLs, links between their answers). Other techniques rely on the co-occurrence frequency of query pairs, which are part of the same search mission [5,7–9]. However, these approaches do not take into account the user's search goal. A recent attempt to tackle this issue is presented in [10].

In [10], the authors propose a general approach to context-aware search using a variable length hidden Markov model (vlHMM). This work is motivated by the belief that the context of a users query, i.e. the past queries and clicks in the same session, may help understand the users information need and improve the search experience substantially. Cao et al. [10] develop a strategy for parameter initialization within the vlHMM learning, which can reduce the number of parameters to be estimated in practice. Additionally, they devise a method for distributed vlHMM learning under the map-reduce model. Within this context, the authors also argue that by considering only correlations between query pairs, the model cannot capture well the users search context. In order to achieve general context-aware search, a comprehensive model is needed that can be used simultaneously for multiple applications (e.g. query suggestion, URL recommendation, document re-ranking). They propose a novel model to support context-aware search and develop efficient algorithms and strategies for learning a very large vlHMM from big log data. The experimental results show that this vlHMM-based context-aware approach is effective and efficient.

Despite the extensive research in this area and the successful application of such methods, they are not suitable for QA systems for at least two reasons. First, the recommendation items are represented by questions as well-formed grammatical units endowed with semantic content, whereas search queries are usually a collection of keywords. Secondly, most QA systems are used with the purpose of learning (e.g., find an explanation for a particular phenomenon, understand a specific concept, etc.), while search engines are usually queried to simply retrieve information.

The work presented in this paper aims at improving question recommendation for QA systems by addressing these two aspects. Our main objective is to leverage the functionality of QA systems towards new learning techniques and use the wisdom of the crowds in order to convey useful information and guide the learner on a meaningful learning journey. For this purpose, a domain-specific topic-taxonomy and Bloom's learning framework is employed, whereas for modeling the order in which questions are selected, variable length Markov chains (VLMC) [6] are used.

The rest of the paper is structured as follows: Section 2 presents the knowledge-base with the domain-specific and learning taxonomies; Section 3 introduces the new learning-oriented recommender model; Section 4 gives an overview of the evaluation results and, finally, Section 5 makes a summary, draws some important conclusions and presents future work objectives.

2   Approach

Aiming at improving the learning experience of users when interacting with a QA system, question recommendation, in the context of this paper, refers to recommending questions to users who ask them and are interested in learning about a particular domain. The question-answer dialog with the system should allow the user to navigate through a meaningful and useful chain of answers that can enrich the users knowledge about a particular domain.

In order to account for the learning process or the order in which questions are selected, a probabilistic graphical model based on variable length Markov chains [6] is constructed and trained on the users question browsing history. This is not a novel approach; it has been successfully adopted for query recommendation [10] as well. In this paper, we attempt to adapt and improve this approach for question recommendation by considering two relevant question features: the questions topic (or subject) and learning objective. The learning objectives, in the context of Blooms learning taxonomy [2], refer to a classification of educational goals (e.g. summarizing, classifying, recognizing, etc.). Current conceptions about learning assume learners as active agents and not passive recipients or simple recorders of information. This shift away from a passive perspective on learning towards more cognitive and constructionist perspectives emphasizes what learners know (knowledge) and how they think (cognitive processes) about what they know [2]. Therefore, the learning taxonomy is defined based on two dimensions: the knowledge and the cognitive process.

For this purpose, two taxonomies are considered: a domain-specific taxonomy that contains possible question topics and Blooms learning taxonomy, a classification of existing learning objectives. In the following, we will present the knowledge base behind the recommendation model.

2.1 Knowledge Base

Let $\mathcal{Q}$ be a set of questions from a particular domain. In general, QA systems can cover several domains of interest, but, for simplicity, we will consider in the following only a single domain.

The Domain-Specific Taxonomy. Let $\mathcal{T}$ be a set of predefined topics from a particular domain. In general, $|\mathcal{Q}| \gg |\mathcal{T}|$. The structure of the corresponding topic taxonomy is given by a generalization-specification relationship between topics:

$$\mathcal{P} \subseteq \mathcal{T} \times \mathcal{T}, (\tau_i, \tau_j) \in \mathcal{P} \iff \tau_i \ parent \ of \ \tau_j, \tag{1}$$

where $\tau_i, \tau_j \in \mathcal{T}$.

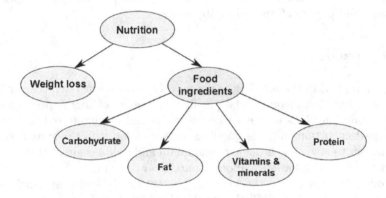

Fig. 1. Snapshot of an example nutrition taxonomy

Consider also a mapping relationship between $\mathcal{Q}$ and $\mathcal{T}$, which maps to each question $q \in \mathcal{Q}$ a topic $\tau \in \mathcal{T}$. A mapping (q, τ) has the following meaning: "question q is about topic τ". We will further refer to this relationship as topic mapping and it is defined in the following way:

$$\mathcal{M}_\tau \subseteq \mathcal{Q} \times \mathcal{T}, (q, \tau) \in \mathcal{M}_\tau \iff q \ is \ mapped \ to \ topic \ \tau. \tag{2}$$

The topic mapping $\mathcal{M}_\tau$ is a surjection with respect to Q (i.e. all questions are mapped to at least one topic). Moreover, one question can be mapped to several topics and one topic can map several questions (see Figure 2).

The Learning Taxonomy. Additionally, we enrich the knowledge base with learning objectives, as defined in Bloom's taxonomy [2]. Let $\mathcal{L}$ be the set of all learning objectives. Every learning objective $\phi \in \mathcal{L}$ is defined as a pair of knowledge and cognitive process instances $(\kappa, \rho) \in \mathcal{K} \times \mathcal{C}$, where

$$\mathcal{K} = \{factual, \ conceptual, \ procedural, \ metacognitive\} \tag{3}$$

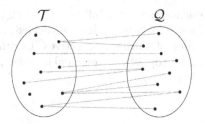

Fig. 2. Question-topic mapping

is the knowledge dimension and

$$\mathcal{C} = \{remember,\ understand,\ apply,\ analyze,\ evaluate,\ create\} \quad (4)$$

is the cognitive process dimension. A more detailed explanation of these concepts can be found in [2]. Similarly to the topic mapping, we define the learning objective mapping as

$$\mathcal{M}_\phi \subseteq \mathcal{Q} \times \mathcal{L}, (q, \phi) \in \mathcal{M}_\phi \iff q\ is\ mapped\ to\ learning\ objective\ \phi. \quad (5)$$

In contrast to the topic mapping $\mathcal{M}_\tau$, we allow questions to be mapped to a single learning objective. Intuitively, this means that a question can refer to several topics, but a single learning goal. In general, questions refer also to a single topic. For simplicity, we have only dealt with single topic assignments in our experiments and, in the following, we consider $\mathcal{M}_{t}au$ to map each question to a single topic.

Question Projections. Based on the mapping relationships, we define the following projection functions:

1. The **topic projection** - a function that projects a question on the topic space using the mapping $\mathcal{M}_\tau$:

$$p_\tau : \mathcal{Q} \to \mathcal{T}, p_\tau(q) = \tau \iff \exists(q, \tau) \in \mathcal{M}_\tau \quad (6)$$

2. The **learning objective projection** - a function that projects a question on the learning objective space using the mapping $\mathcal{M}_\phi$:

$$p_\phi : \mathcal{Q} \to \mathcal{L}, p_\phi(q) = \phi \iff \exists(q, \phi) \in \mathcal{M}_\phi \quad (7)$$

2.2   The Learning-Oriented Recommendation Model

In the following, a novel learning-oriented question recommendation technique is introduced that aims at improving the user's learning experience based on the following intuition.

Intuition: Question sequences are first influenced by the underlying topics or subject and the order in which these topics are tackled, and then, within each topic, by a particular order of learning objectives. In other words, users tend to ask questions grouped by topics; in a particular order given by the question learning objectives (see Figure 3).

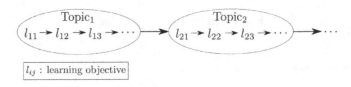

Fig. 3. Intuition behind the user learning process

This intuition emerged during the evaluation process, where several probabilistic recommendation models were constructed and tested. The results showed that the model based on this intuition performed better than the rest of them. Due to the limited space, only three of the most relevant models will be considered here for comparison.

Preliminaries. Let Q, T and L be random variables taking values in the question set $\mathcal{Q}$, the topic space $\mathcal{T}$ and the set of learning objectives $\mathcal{L}$, respectively.

Consider $\mathcal{H}$ to be the history database which contains, for each user, an ordered sequence of questions representing the user's history log.

A learner is given a training set (usually a subset of the history database $\mathcal{H}$) of question sequences $q_1^n = q_1 q_2 \ldots q_n$, where $q_i \in \mathcal{Q}$ and $q_i q_{i+1}$ means that question q_i was asked before question q_{i+1}.

Given this training set, our goal is to learn a model P that provides a probability assignment for any future outcome given some past. More specifically, given a context of previously asked questions $s \in \mathcal{Q}^*$ (i.e. an ordered sequence of the user's past question selections) and a question q, the learner should generate a conditional probability distribution $P(q|s)$.

We measure the *prediction performance* using the average log-loss [6] $l(P, x_1^t)$ of P with respect to a test sequence $x_1^t = x_1 x_2 \ldots x_t$:

$$l(P, x_1^t) = -\frac{1}{t} \sum_{i=1}^{t} \log(P(x_i | x_1 \ldots x_{i-1})) \qquad (8)$$

where x_i represent questions in $\mathcal{Q}$. The average log-loss is directly related to the likelihood $P(x_1^t) = \prod_{i=1}^{t} P(x_i | x_1 \ldots x_{i-1})$ and, therefore, minimizing the average log-loss is equivalent to maximizing the likelihood.

The Recommendation Model. Based on the learned model P, we define the *learning-oriented recommender* (LoR). For each user with history log $s \in \mathcal{Q}^*$, we want to recommend a set $R(s) \subseteq \mathcal{Q}$ of N questions that satisfies the following:

$$R(s) = \underset{q \in \mathcal{Q} \text{ and } q \notin s}{\arg\max^N} \; P(q|s) \qquad (9)$$

where $\arg\max^N$ returns the first N maximal arguments with respect to the given function. In other words, the learning oriented recommender tries to recommend the first N best questions that maximize the user's utility. In this case, the utility is dependent on the learned model P.

P is learned according to a probabilistic graphical model based on hidden VLMCs: one with hidden states T and then, for each $\tau \in T$ a VLMC with hidden states L. The observation states are given by Q over the question space $\mathcal{Q}$ (see Figure 4).

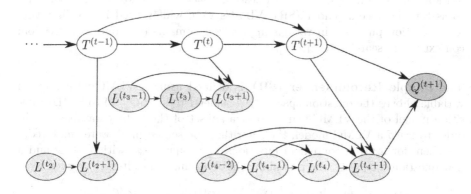

Fig. 4. The learning-oriented recommender

To learn such a model, first, the training sequences are projected on the topic space using p_τ and a VLMC over T is trained on them. As a result, the transition model $P(T^{(t+1)}|T^{(1:t)})$ is obtained.

Then, for each topic τ, a transition probability $P_\tau(L^{(t'+1)}|L^{(1:t')})$ is learned by training a VLMC over L on the projections of the question sub-sequences within topic τ, using the learning objective projection function p_ϕ.

We define the observation model $P(Q^{(t+1)}|T^{(t+1)}, L^{(t+1)}, Q^{(1:t)})$ as the probability of randomly sampling an unvisited question corresponding to topic $T^{(t+1)} = \tau_{t+1}$ and learning objective $L^{(t+1)} = \phi_{t+1}$

$$P(q_{t+1}|\tau_{t+1}, \phi_{t+1}, q_1^t) = \begin{cases} 0 & \text{if } \nexists(q_{t+1}, \tau_{t+1}) \in \mathcal{M}_\tau \lor \nexists(q_{t+1}, \phi_{t+1}) \in \mathcal{M}_\phi \\ \frac{1}{S} & \text{otherwise} \end{cases},$$

$$(10)$$

where $S = \{q' \in \mathcal{Q} \backslash \{q_1, \ldots, q_t\} | (q', \tau_{t+1}) \in \mathcal{M}_\tau \land (q', \phi_{t+1}) \in \mathcal{M}_\phi\}$.

Prediction. Let $x_1^t = x_1 \ldots x_t$ with $x_i \in \mathcal{Q}, i \in \{1, \ldots, t\}$ be a context sequence of questions and $x_{t+1} \in \mathcal{Q}$ be the user's next question. Then, we define the probability of observing question x_{t+1} after x_1^t as:

$$P(x_{t+1}|x_1^t) = P(\tau_{t+1}|\tau_1^t) \cdot P(\phi_{t+1}|\phi_1^t) \cdot P(x_{t+1}|\tau_{t+1}, \phi_{t+1}, x_1^t), \qquad (11)$$

where $\tau_{t+1} = p_\tau(x_{t+1})$ is the projection of question x_{t+1} on the topic space $\mathcal{T}$, $\phi_{t+1} = p_\phi(x_{t+1})$ is the projection of question q_{t+1} on the space of learning objectives $\mathcal{L}$, $\tau_1^t = p_\tau(x_1^t) = p_\tau(x_1) \ldots p_\tau(x_t)$ and $\phi_1^t = p_\phi(x_1^t) = p_\phi(x_1) \ldots p_\phi(x_t)$, x_1^t being the last sub-sequence within topic τ.

2.3 Other Models

In order to show the effectiveness of the introduced learning-oriented question recommender, we intend to perform a comparison with other models that use Markov chains. One of the most widely used ones is a simple variable length Markov chain (VLMC) over the question space, which we will further refer to as Simple Recommender (SR). VLMCs were widely used in the literature for prediction purposes in various application domains (e.g. data compression, context-aware search, etc.) [6,10].

The Simple Recommender (SR) is defined using a VLMC with random variable Q over the question space $\mathcal{Q}$. Consider $P(Q^{(t+1)}|Q^{(1:t)})$ to be the transition model of the VLMC trained over a subset of the history database $\mathcal{H}$. In order to learn a VLMC model, the algorithms presented in [6] were employed.

Then, for a given context of questions $x_1^t = x_1 x_2 \cdots x_t$ with $x_i \in \mathcal{Q}$ and a new question $x_{t+1} \in \mathcal{Q}$, the probability of observing x_{t+1} after x_1^t is given by:

$$\begin{aligned}
P(x_{t+1}|x_1^t) &= P(Q(t+1) = x_{t+1}|Q^{(1)} = x_1, \ldots, Q^{(t)} = x_t) \\
&= P(Q^{(t+1)} = x_{t+1}|Q^{(t\lambda+1)} = x_{t\lambda+1}, \ldots, Q^{(t)} = x_t),
\end{aligned} \qquad (12)$$

where $\lambda = \lambda(x_t, x_{t1}, \ldots)$ is a function of the past determined during the learning process of the VLMC. Let $D = \max_{x_t, x_{t1}, \ldots} \lambda(x_t, x_{t1}, \ldots)$ be the maximal memory length of the VLMC. Figure 5 represents a simple recommender model with $D = 3$.

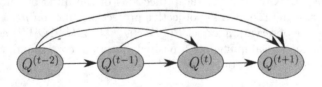

Fig. 5. Simple question recommender based on a VLMC

Intuition: The SR model is based on the intuition that the unique question identifiers are enough to efficiently identify learning patterns within question sequences and use them to produce accurate predictions.

The Random Recommender (RR) is a model that simply recommends questions randomly, without relying on any knowledge or user history log when producing recommendations. This model was considered only as base comparison to show that the previous two recommendation models (LoR and SR) are not random and that they actually outperform greatly the RR model.

The following section will show the performance of each of these recommendation models on three different knowledge bases.

3   Evaluation and Results

3.1   Data

In order to thoroughly test the performance of the LoR model, three data sets of questions, corresponding to three different domains, were collected: earth sciences (from Wiki Answers[1] and MadSci[2]), nutrition (provided by Sasha Walleczek[3]) and homeschooling (from Wiki Answers).

For reasons of robustness, corresponding to each of these data sets, a topic taxonomy with the structure presented in Section 2.1 was manually constructed. Table 1 gives an overview on the size of the data sets and the corresponding taxonomies. None of these taxonomies reflect a unique and complete image of the actual domains. They are merely a snapshot of the domains from a particular perspective. The topic-trees were constructed in a way to cover the question datasets. In this particular case, the structure of the hierarchy does not influence the performance of the recommender models and, therefore, represents no variable in the overall evaluation process. However, the performance of the recommender models does depend on the topic, knowledge and cognitive process mappings.

Table 1. Overview of question data sets

Data set	No. of questions	No. of topics	No. of questions with questions
Earth sciences	313	49	37
Nutrition	318	38	24
Homeschooling	191	42	39

---

[1] www.wiki.answers.com

[2] www.madsci.org

[3] www.walleczek.at

Similarly, the assignment of questions to the set of topics $\mathcal{T}$ and learning objectives $\mathcal{L}$ was performed manually in order to maintain robustness. However, not all topics or learning objectives were identified within the three question sets. Table 2 gives an overview of the mappings' statistics. Column $avg_{\tau' \in \mathcal{T}}(|\{(q, \tau') \in \mathcal{M}_\tau\}|)$ contains the average number of question per topic, while $avg_{\phi' \in \mathcal{L}}(|\{(q, \phi') \in \mathcal{M}_\phi\}|)$ represents he average number of question per learning objective (12 in total).

Table 2. Statistics on the topic and learning objective mappings

Data set	$avg_{\tau' \in \mathcal{T}}(|\{(q, \tau') \in \mathcal{M}_\tau\}|)$	$avg_{\phi' \in \mathcal{L}}(|\{(q, \phi') \in \mathcal{M}_\phi\}|)$
Earth sciences	8.46	34.78
Nutrition	13.25	26.5
Homeschooling	4.89	19.1

3.2 Experiment

The evaluation of the recommender models introduced in Subsections 2.2 and 2.3 is not an easy task for several reasons. First, to learn such models, a history of user interactions with the QA system is needed. Without any kind of recommendation engine behind the search or browsing functionality, such interactions would not be possible, or even reliable, since the user is not aware of the possible question choices.

Secondly, if suggestions are provided, even in their simplest form, the resulting browsing log would not reflect the users natural learning process, but rather a learning process influenced by the capabilities of the used recommendation engine. Therefore, the recorded question sequences would still not be suitable to be used for training a new recommender model which relies on the natural learning process of the user.

In order to evaluate the performance of the LoR, due to the lack of resources, a scenario of user interaction with a QA system was simulated, where recommendations were not provided at all. Having an overview of the available questions is not feasible, given the size of the datasets. Therefore, for each domain, five subsets of 20 questions were randomly generated and users were asked to order each of these 20 question-sets in the sequence that they, personally, would want to ask them or would want **learn** about.

Table 4 shows, for each of the three domains, the number of collected question sequences, i.e. the total number of user responses, the number of distinct questions within the collected sequences and their percentage with respect to the total number of questions.

Overall, about 13 male and female users participated in this survey, but not all of them provided an ordering for each of the question sets. The obtained number of sequences are generally balanced between male and female participants.

Table 3. Statistics of the collected sequences

Data set	No. of sequences	No. of distinct questions	%
Earth sciences	61	90	28.75
Nutrition	46	94	29.56
Homeschooling	56	84	43.98

3.3 Evaluation Metrics

Evaluating a recommender system on its prediction power is crucial, but insufficient in order to deploy a good recommendation engine [17]. There are other measures that reflect various aspects. However, not all of them are desired to perform well for every recommender.

Therefore, the evaluation of the LoR model should not be based on prediction performance (accuracy and average log-loss) alone, but also on other metrics that capture various desired aspects of a learning-oriented recommender within a QA system. Let us briefly define these metrics.

Catalog Coverage. In general, catalog coverage represents the proportion of questions that the recommendation model can recommend. In our case, we define the catalog coverage as the proportion of questions that the model P can recommend with a prediction value higher than a predefined threshold σ.

Overall, all three recommender models introduced in Section 2 can generate recommendations for any user (i.e. full user space coverage) and, eventually, all questions can be recommended, since the recommender repeatedly excludes already visited ones. But, towards the exhaustion of the database, the recommendations will have a very low prediction value. These recommendations are unreliable. Therefore, we introduce the prediction threshold σ.

In our evaluation, we generally set σ to be the lowest prediction value among the questions within the sequences used for training. Since the user space coverage is equal for all recommender models, we will further refer to catalog coverage simply as "coverage".

Diversity. Generally, diversity is defined as the opposite of similarity. Within this context, we define the diversity as the average dissimilarity among each question pair within a recommendation.

Let s be a question sequence context. Then, the diversity of $R(s)$ is defined as

$$div(R(s)) = \frac{2}{N \cdot (N-1)} \sum_{\substack{(q_i, q_j) \in R(s) \\ i < j}} [1 - sim_q(q_i, q_j)], \qquad (13)$$

where $sim_q : \mathcal{Q} \times \mathcal{Q} \to [0, 1]$ represents the semantic similarity measure between questions.

During the evaluation, we used the simple cosine similarity together with the semantic concept similarity defined by Lin [14]. In order to avoid further dependencies with our topic taxonomy, the Wordnet [18] lexical database was used instead

Learning Utility. The learning utility refers to the learning gain of a user from a recommendation. One way of measuring learning utility is with user ratings. Since such an experiment can only be performed within a user study setting, a comparative metric is introduced instead that shows how good a model reflects the user learning process.

Consider two sets of equal size: S_{learn} a set of user question sequences based on the user's learning process (like the ones collected during our experiment) and S_{rand} a set of randomly generated question sequences. Each of the sequence pairs from $S_{learn} \times S_{rand}$, corresponding to the same user, have the same length. Now let M be a recommendation model. We train this model with each of the two sequence sets using cross-validation and obtain the accuracy values:

$$a_{learn} = acc(M, S_{learn}) \quad \text{and} \quad a_{rand} = acc(M, S_{rand}). \qquad (14)$$

We define the learning utility of model P by comparing the normalized accuracy difference:

$$lu(P, S_{learn}, S_{rand}) == \begin{cases} 0 & if \ a_{learn} = 0 \\ \frac{a_{learn}}{a_{learn} - a_{rand}} & \text{otherwise} \end{cases}. \qquad (15)$$

This measure works only under the assumption that the set S_{learn} truly reflects the users' learning process. It shows how dependent model P is on receiving as input learning sequences.

3.4   Results

In the first part, an analysis of the survey results was made, in order to have an overview of the generated sequences and to identify early patterns and correlations between user answers. The results show that, in some cases, the users strongly agree on a particular question sequence, yet in other cases major discrepancies were identified (see Figure 6). This can be explained by the unique and personal way humans understand certain concepts, i.e. the unique conceptual world map existing in each human mind. Additionally, some domain-specific questions are rather ambiguous and up for interpretation. The survey also captures user preferences and personal opinions and, therefore, there are no unanimous answers. For our evaluation purposes, this aspect was preferred over highly correlated question sequences, because it reflects real life situations. Hence, the learned models are not highly accurate, but despite the conflicting user opinions, some of them still proved to identify learning process patterns and use them to make useful recommendations.

In order to show the benefits of the learning-oriented recommender, two other models were considered: a simple recommender (SR) using a VLMC of random variable Q over the question space and a random recommender (RR)

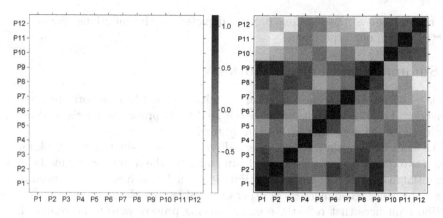

Fig. 6. Correlation matrices of user question orderings for the earth sciences domain

that recommends questions randomly. The SR corresponds to the approach proposed in [10]. Table 4 shows the results obtained using a 10-fold cross validation with input parameters: number of recommendations $N = 5$, maximum order of the VLMCs, $maxOrder = 10$ and $\sigma = $ the lowest prediction value among the questions within the sequences used for training. For testing, the leave-one-out technique was employed.

Table 4. Results

Data set	Model	acc	avg-ll	cov	div	lu
Earth sciences	SR	0.65	93.84	0.29	0.47	0.45
	LoR	0.31	137.69	**0.60**	0.44	**0.69**
	RR	0.01	164.9	1	0.60	0
Nutrition	SR	0.52	112.34	0.30	0.50	0.41
	LoR	0.15	156.23	**0.67**	0.51	**0.49**
	RR	0	165.38	1	0.40	0
Homeschooling	SR	0.50	107.87	0.44	0.36	0.39
	LoR	0.16	145.92	**0.84**	0.47	**0.41**
	RR	0.06	150.06	1	0.42	0

Although the LoR did not achieve an accuracy (acc) and average log-loss ($avg - ll$) as high as the SR, compared to the RR, it still had a good prediction performance (see Table 4). However, the coverage (cov) and learning utility (lu) values of the LoR were much higher, whereas the diversity did not show significant discrepancies among the three models. The coverage values show that the LoR, compared to the SR, can recommend a larger percentage of questions. The increased learning utility of the LoR shows that the prediction performance of this recommendation model is more dependent on receiving as input learning sequences, like the ones collected during our experiments. This means that the

LoR reflects better the learning process depicted in the questions sequences depicted during our experiment.

4   Summary and Future Work

In this paper, a new recommendation technique, called learning-oriented recommender (LoR), is introduced with the goal to improve the user's learning experience while interacting with a QA system.

The evaluation of the learning-oriented recommender is not as easy task for at least two reasons. First, in order to train and learn the recommender model, a substantial history of user learning activity is needed, which is not influenced in any way by other recommenders or other external factors. Secondly, even if such question sequences that reflect the users learning process were to be collected, there is no clear, well established metric to evaluate the performance of the recommender from a learning perspective.

However, a first step was made towards a better understanding of the learning-oriented recommender's capabilities. From each of the above mentioned three datasets of questions (i.e. earth sciences, nutrition and homeschooling), five sets of 20 questions were randomly selected and users were asked to order each set according to their learning preferences in the sequence that they, personally, would ask them or want to learn about.

Five evaluation measures were used to compare the performance of the LoR with a simple VLMC (SR) over the question sequences and a random recommender (RR). Results show that while the SR outperforms the LoR with respect to prediction power, the LoR achieved a much higher coverage and learning utility. The RR was used as a base reference. The obtained results confirm our initial intuition: question sequences are first influenced by the underlying topics and their order, and then, within each topic, by a particular order of learning objectives.

However, further evaluation is required to show that the LoR has great potential in offering an improved user learning experience. To show this in more detail, we intend to conduct an online user study. Additionally, we also plan to analyze the influence of the knowledge-based structure on the recommendation performance.

Additionally, it would be desirable to investigate the potential of an automatic topic-tree generation, and, more importantly the automatic assignment of questions to topics and to learning objectives.

With the information overload, new aspects of existing disciplines are identified or entirely unknown, unexplored fields of study are discovered. In the first case, a restructuring or extension of the current curriculum is required. The second case demands the settlement of the first building blocks. Learning patterns represent relevant knowledge about these domains. By using the learning patterns derived from our recommender model, we could establish new fields of study and (semi-)automatically generate curricula for these domains. Further research in this direction is expected to answer the question whether and how to exploit the learning-oriented recommender model for this purpose.

References

1. Adomavicius, G., Tuzhilin, A.: Toward the Next Generation of Recommender Systems: A Survey of the State-of-the-Art and Possible Extensions. IEEE Trans. on Knowl. and Data Eng. **17**(6), 734–749 (2005)
2. Anderson, L., Krathwohl, D., Airasian, P., Cruikshank, K., Mayer, R., Pintrich, P., Raths, J., Wittrock, M. (eds.): A taxonomy for learning, teaching, and assessing: A revision of Bloom's taxonomy of educational objectives. Addison Wesley Longman Inc. (2001)
3. Baeza-Yates, R.: Graphs from search engine queries. In: van Leeuwen, J., Italiano, G.F., van der Hoek, W., Meinel, C., Sack, H., Plášil, F. (eds.) SOFSEM 2007. LNCS, vol. 4362, pp. 1–8. Springer, Heidelberg (2007)
4. Baeza-Yates, R., Hurtado, C.A., Mendoza, M.: Query recommendation using query logs in search engines. In: Lindner, W., Fischer, F., Türker, C., Tzitzikas, Y., Vakali, A.I. (eds.) EDBT 2004. LNCS, vol. 3268, pp. 588–596. Springer, Heidelberg (2004)
5. Bai, L., Guo, J., Cheng, X.: Query Recommendation by Modelling the Query-Flow Graph. In: Salem, M.V.M., Shaalan, K., Oroumchian, F., Shakery, A., Khelalfa, H. (eds.) AIRS 2011. LNCS, vol. 7097, pp. 137–146. Springer, Heidelberg (2011)
6. Begleiter, R., El-Yaniv, R., Yona, G.: On prediction using variable order Markov models. Journal of Artificial Intelligence Research **22**, 385–421 (2004)
7. Boldi, P., Bonchi, F., Castillo, C., Donato, D., Gionis, A., Vigna, S.: The query-flow graph: model and applications. In: Proceedings of the 17th ACM Conference on Information and Knowledge Management, CIKM 2008, pp. 609–618. ACM, New York (2008)
8. Boldi, P., Bonchi, F., Castillo, C., Donato, D., Vigna, S.: Query suggestions using query-flow graphs. In: Proceedings of the 2009 Workshop on Web Search Click Data, WSCD 2009, pp. 56–63. ACM, New York (2009)
9. Bonchi, F., Perego, R., Silvestri, F., Vahabi, H., Venturini, R.: Recommendations for the long tail by term-query graph. In: Proceedings of the 20th International Conference Companion on World Wide Web, WWW 2011 (2011)
10. Cao, H., Jiang, D., Pei, J., Chen, E., Li, H.: Towards context-aware search by learning a very large variable length hidden Markov model from search logs. In: Proceedings of the 18th International Conference on World Wide Web, WWW 2009, pp. 191–200. ACM, New York (2009)
11. Hu, D., Gu, S., Wang, S., Wenyin, L., Chen, E.: Question recommendation for user-interactive question answering systems. In: Proceedings of the 2nd International Conference on Ubiquitous Information Management and Communication, ICUIMC 2008, pp. 39–44. ACM, New York (2008)
12. Kabutoya, Y., Iwata, T., Shiohara, H., Fujimura, K.: Effective Question Recommendation Based on Multiple Features for Question Answering Communities. In: Proceedings of the Fourth International Conference on Weblogs and Social Media, ICWSM 2010, Washington, DC, USA, May 23–26 (2010)
13. Kosorus, H., Bgl, A., Kng, J.: Semantic similarity between queries in a QA system using a domain-specific taxonomy. In: Proceedings of the 14th International Conference on Enterprise Information Systems, pp. 241–246, Wrocław, Poland (June 2012)

14. Lin, D.: An information-theoretic definition of similarity. In: Proceedings of the 15th International Conference on Machine Learning, pp. 296–304 (1998)
15. Qu, M., Qiu, G., He, X., Zhang, C., Wu, H., Bu, J., Chen, C.: Probabilistic question recommendation for question answering communities. In: Proceedings of the 18th International Conference on World Wide Web, WWW 2009, pp. 1229–1230. ACM, New York (2009)
16. Santos, O., Boticario, J.G. (eds.): Educational Recommender Systems and Technologies: Practices and Challenges. IGI Global (2011)
17. Shani, G., Gunawardana, A.: Evaluating recommendation systems. In: Ricci, F., et al. (eds.) Recommender Systems Handbook. Springer Science+Business Media, LLC (2011)
18. Princeton University. Wordnet. www.wordnet.princeton.edu (2010)

On the Performance of Triangulation-Based Multiple Shooting Method for 2D Geometric Shortest Path Problems

Phan Thanh An[1,2], Nguyen Ngoc Hai[3], Tran Van Hoai[4],
and Le Hong Trang[1,5](✉)

[1] Instituto Superior Técnico, CEMAT, Av. Rovisco Pais,
1049-001 Lisboa, Portugal
lhtrang@math.ist.utl.pt
[2] Institute of Mathematics, 18 Hoang Quoc Viet road, 10307 Hanoi, Vietnam
[3] Department of Mathematics, International University,
Vietnam National University, Thu Duc, Ho Chi Minh City, Vietnam
[4] Faculty of Computer Science and Engineering, HCMC University of Technology,
268 Ly Thuong Kiet Street, Ho Chi Minh City, Vietnam
[5] Faculty of Information Technology, Vinh University, 182 Le Duan,
Vinh, Vietnam

Abstract. In this paper we describe an algorithm based on the idea
of the direct multiple shooting method for solving approximately 2D
geometric shortest path problems (introduced by An et al. in Journal
of Computational and Applied Mathematics, 244 (2103), pp. 67-76).
The algorithm divides the problem into suitable sub-problems, and then
solves iteratively sub-problems. A so-called collinear condition for combining the sub-problems was constructed to obtain an approximate solution of the original problem. We discuss here the performance of the
algorithm. In order to solve the sub-problems, a triangulation-based algorithm is used. The algorithms are implemented by C++ code. Numerical
tests for An et al.'s algorithm are given to show that it runs significantly
in terms of run time and memory usage.

Keywords: Approximate algorithm · Multiple shooting method · Memory usage · Run time · Shortest path

1 Introduction

The problem of finding the Euclidean shortest path between two points in a
simple polygon is one of fundamental problems in computational geometry and
arises in many applications such as graph algorithms, geographic information system (GIS), robotics, etc. (see [3]). There are many approximation algorithms for

An erratum to this chapter is available at DOI 10.1007/978-3-662-45947-8_8

© Springer-Verlag Berlin Heidelberg 2014
A. Hameurlain et al. (Eds.): TLDKS XVI, LNCS 8960, pp. 45–56, 2015.
DOI: 10.1007/978-3-662-45947-8_4

solving shortest path problems in which the algorithms are constructed relying on graph theory (see [3,6]). The main challenge of these methods is concentrated in the size of the problem. Namely when the size of the problem is huge, the required resource for computing is thus increased too much. This can be overcome ideally by dividing the problem into sub-problems. The sub-problems are solved. The obtained shortest paths of the sub-problems are then combined into the shortest path of the original problem.

For solving ordinary differential equations (ODE), in particular with boundary value problems, the multiple shooting method was introduced (see [13]). By this method, the problem is first divided into several sub-problems. The values of the solutions of sub-problems are computed simultaneously by iteration. To this end, a continuity condition between the solutions of the sub-problems is required at the solution of the problem. A variant of the multiple shooting method called direct multiple shooting method was introduced by Bock and Plitt [4] to give the solution of optimal control problems. Based on the idea of this method, an approximate algorithm for solving 2D geometric shortest path problems has been recently introduced by An et al. [1]. Given a simple polygon, the algorithm first divides the polygon into sub-polygons then computes iteratively to obtain an approximate shortest path between two vertices of the polygon. A so-called collinear condition based on the idea of continuity and differentiability conditions for combining the solutions in the sub-polygons, is constructed to obtain an appropriate shortest path in the polygon (see Theorem 3.2-3.3 in [1]).

In this paper, we discuss in detail the algorithm introduced in [1] on the aspect of performance. It differs from [1] that we implement a triangulation-based algorithm called the funnel algorithm (see [7,8,10]) for finding the geometric shortest path in each sub-polygon instead of using the algorithm given in [2]. The funnel algorithm is based on triangulation of the polygon and graph theory. The numerical tests show that the method is efficient and reliable.

2   Description of Multiple Shooting Method for Solving the 2D Shortest Path Problems

2.1   Multiple Shooting Method for Solving ODE-Boundary Value Problems

Consider the boundary value problem of seeking a differentiable function $y = f(x)$ of one variable x such that its derivative $y'(x)$ satisfies

$$y'(x) = f(x,y), \tag{1}$$
$$r\big(y(a), y(b)\big) = 0, \tag{2}$$

where (2) specifies the boundary condition and a, b are two different numbers.

In a multiple shooting method (see [13]), the domain of solution of the problem is discretized into several intervals as follows,

$$a = x_1 < x_2 < \cdots < x_k = b.$$

The values $s_i = y(x_i)$ of the solution $y(x)$ of the problem are computed by iteration, for $i = 1, \ldots, k$. In order to obtain the solution of the original problem, a solution of the initial-value problem

$$y' = f(x, y), \text{ for } x \in [x_i, x_{i+1}),$$
$$y(x_i) = s_i,$$

is determined, denoted by $y(x; x_i, s_i)$, for $i = 1, \ldots, k$. Then by the continuity of $y(x)$, the problem now is to determine s_i, for $i = 1, \ldots, k$, such that the following conditions are satisfied (see Fig. 1),

$$s_{i+1} = y(x_{i+1}; x_i, s_i), \text{ for } i = 1, \ldots, k-1,$$
$$y(b) = s_k,$$
$$r(s_1, s_k) = 0.$$

Because $y(x)$ is differentiable on $[a, b]$, if the continuity condition is satisfied at x_i, then

$$y'_-(x_{i+1}; x_i, s_i) = y'_+(x_{i+1}; x_{i+1}, s_{i+1}),$$

where $y'_-(x_i)$ and $y'_+(x_i)$ denote the left and right derivatives of $y(x)$ at x_i, respectively, for $i = 1, \ldots, k-2$. This means that the left and right tangent lines to the graph of $y(x)$ are collinear at the exact solution point (x_i, s_i^*) (see Fig. 1).

The idea of continuity and differentiability of the multiple shooting method leads us to define a so-called collinear condition in Subsection 2.3.

2.2 A Triangulation-Based Algorithm for Finding Shortest Paths in Sub-polygons

Algorithm given in [1] is based on the idea of the direct multiple shooting method which is a variant of the multiple shooting approach for solving optimal control problems [4]. The algorithm divides a polygon into suitable sub-polygons. A procedure is called to determine the shortest paths in the sub-polygons. An algorithm based on incremental convex hull was used for the procedure. In this paper, in order to analyze the performance of the algorithm we use instead the funnel algorithm (see [7,8,10]). A triangulation of the polygon and graph theory are required for the funnel algorithm. These are crucial factors that effect to the performance of the algorithm for finding the geometric shortest paths.

For the funnel algorithm we first triangulate the polygon. Once the triangulation of polygon is determined, a dual tree of the triangulation is constructed based on the adjacent relationship between triangles of the triangulation. On the dual tree, a sequence of consecutive triangles, i.e., two adjacent triangles of the sequence share an edge, is determined so that the shortest path between starting point a and destination point b is contained inside the sequence. The sequence is said to be a *sleeve*. Each sharing edge of the sequence is said to be a *diagonal* of the sleeve. The funnel algorithm consists of three components: the *path*, the *apex*, and the *funnel* (see Fig. 2).

At the start of the funnel algorithm, the apex is set to the starting point a. The funnel is initialized by two line segments from apex to the end points of the first diagonal. We consider next diagonals of the sleeve to extend the funnel. At some step a new apex can be established, the path connecting old apexes to the new apex is a part of the shortest path between a and b. Once the last diagonal is processed, we add the destination point b to obtain whole shortest path. We choose the funnel algorithm because it is feasible to implement.

2.3 Multiple Shooting Method for Solving the 2D Shortest Path Problems

We now describe briefly the direct multiple shooting method for solving shortest path problems (see [1] for more details). Given a simple polygon $\mathcal{D} = \mathcal{P}\mathcal{Q}$ and two vertices a and b of the polygon in which $\mathcal{P}$ ($\mathcal{Q}$, respectively) is the polyline formed by vertices of the polygon from a to b (b to a, respectively) in counterclockwise order. Without loss of generality we assume that $a_x < b_x$, where v_x denotes the x-coordinate of point v in 2D. Let us denote by $SP(a, b)$ the shortest path between a and b in $\mathcal{D}$. Given u and v in 2D and $0 \leq \lambda \leq 1$, we

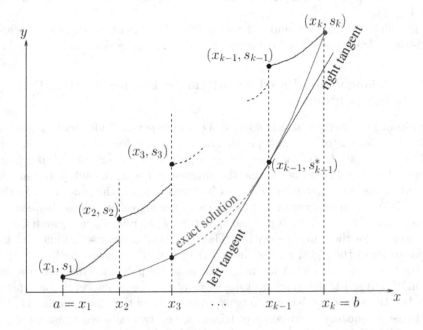

Fig. 1. Multiple shooting for solving ODE-boundary value problem

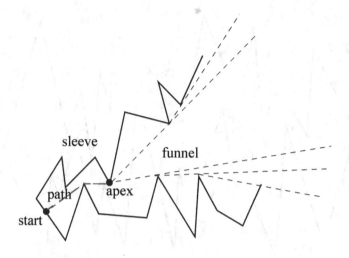

Fig. 2. The path, apex, and funnel of the funnel algorithm

define

$$[u, v] = \lambda u + (1 - \lambda)v,$$
$$[u, v[= [u, v] \setminus \{v\},$$
$$]u, v[= [u, v[\setminus\{u\}.$$

Based on the idea of direct multiple shooting method for solving the optimal control problems, the proposed method first splits the polygon $\mathcal{PQ}$ into sub-polygons $\mathcal{T}_i$, for $i = 0, \ldots, k$, by parallel cutting segments $\xi_1, \ldots, \xi_k$ satisfying following conditions:

$\xi_i = [u_i, v_i] \subset \mathcal{D}$ such that $]u_i, v_i[\subset\text{int}\mathcal{D}$,
$[u_i, v_i]$ strictly separates a and b,
$\mathcal{T}_i$ is bounded by $\mathcal{P}, \mathcal{Q}$, cutting segments ξ_i and ξ_{i+1},

and

$$SP(u_i, u_{i+1}) \cap SP(v_i, v_{i+1}) \neq \emptyset,$$

where $u_i \in \mathcal{P}$ and $v_i \in \mathcal{Q}$, for all $i = 1, \ldots, k$. We then have that $\mathcal{PQ} = \cup_{i=0}^{k}\mathcal{T}_i$ and $\text{int}\mathcal{T}_i \cap \text{int}\mathcal{T}_j = \emptyset$, for $i \neq j$.

We initialize a set of shooting points $a_i^0 \in \xi_i$, for $i = 1, \ldots, k$. The shooting point $a_i^j \in \xi_i$ $(j \geq 0)$ is refined iteratively such that $\mathcal{Z}^j := \cup_{i=0}^{k}\mathcal{Z}_i^j$ tends to $SP(a, b)$, where $a_0^j = a$ and $a_{k+1}^j = b$ and $\mathcal{Z}_i^j$ denotes the shortest path between a_i^j and a_{i+1}^j in $\mathcal{T}_i$ (see Fig. 3). This refinement is finished if the following collinear condition

$$\angle SP(a_i^j, a_{i-1}^j), SP(a_i^j, a_{i+1}^j) = \pi,$$

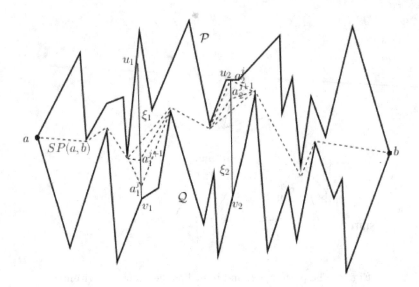

Fig. 3. Given $\mathcal{D} = \mathcal{PQ}$, a and b are vertices of $\mathcal{D}$ where $a_x < b_x$. ξ_1 and ξ_2 are cutting segments where ξ_1 and ξ_2 are parallel to Oy. Shooting points a_1^j and a_2^j are refined such that $\mathcal{Z}^j$ tends to $SP(a, b)$.

is reached. This can be seen as a type of condition of continuity and differentiability in the multiple shooting method for solving the boundary value problems.

The refinement is based on some properties of angles of the shortest path at cutting segments.

Theorem 1 ([1]). *Suppose that a and b are disjoint points of a simple polygon $\mathcal{D}$. $\xi = [u, v]$ is a cutting segment of $\mathcal{D}$ which strictly separates a and b. Let α_u be the angle $\angle SP(u, a), SP(u, b)$ which is an interior angle of $CH_{\mathcal{D}}(a, v, b, u)$ at vertex u and α_v be the angle $\angle SP(v, a), SP(v, b)$ which is an outerior angle of $CH_{\mathcal{D}}(a, v, b, u)$ at vertex v. Then,*

i. $(\pi - \alpha_u)(\pi - \alpha_v) < 0$ iff $SP(a, b)$ intersects $]u, v[$.
ii. If $\alpha_u \geq \pi$ ($\alpha_v \leq \pi$, respectively) then $SP(a, b)$ goes through u (v, respectively) and $\alpha_v > \pi$ ($\alpha_u < \pi$, respectively).

Theorem 2 ([1]). *Suppose that a and b are disjoint points of a simple polygon $\mathcal{D}$. $\xi = [u, v]$ is a cutting segment of $\mathcal{D}$ which strictly separates a and b, and $x \in [u, v]$. Let α_x be the angle $\angle SP(x, a), SP(x, b)$, which is an interior angle of $CH_{\mathcal{D}}(a, v, b, x)$ at vertex x and $SP(a, b)$ intersects $[u, v]$ at z, and $\alpha_u < \pi$. Then α_x is monotone on $[u, z]$ and $[z, v]$.*

The following procedure is used to find the shortest path between a_i^j and a_{i+1}^j, for $i = 0, \ldots, k$, where $a_0^j = a$ and $a_{k+1}^j = b$.

   procedure SP(x, y, ξ, μ)

Find the shortest path between x and y in the polygon bounded by $\mathcal{P}, \mathcal{Q}$, and cutting segments ξ and μ in which $x \in \xi$ and $y \in \mu$.
    end procedure
As mentioned, in [1] a shortest path algorithm based on incremental convex hull was used. We, however, use alternatively the algorithm described in subsection 2.2. Following is the algorithm, based on the idea of the multiple shooting method, for solving approximately the geometric shortest path problem in a simple polygon.

Algorithm 1 *Multiple Shooting Method for 2D Shortest Path Problems*

Input: vertices a and b of a simple polygon $\mathcal{PQ}$, where $\mathcal{P}$ ($\mathcal{Q}$, respectively) is the polyline formed by vertices of the polygon from a to b (b to a, respectively) in counterclockwise order.
Output: an approximate shortest path $\mathcal{Z}$ between a and b in $\mathcal{PQ}$.

1. Divide the polygon $\mathcal{PQ}$ into suitable $\mathcal{T}_i$ by cutting segments $\xi_1, \ldots, \xi_k$. Set $j = 0$. Choose initial points $a_i^j \in \xi_i$, for $i = 1, \ldots, k$.
2. Call SP($a_i^j, a_{i+1}^j, \xi_i, \xi_{i+1}$) to get shortest path $\mathcal{Z}_i^j$ between a_i^j and a_{i+1}^j in $\mathcal{T}_i$. Check if all $\mathcal{Z}_i^j$ satisfy simultaneously the multiple shooting structure, i.e., the collinear condition holds true, then **goto 4**.

    Otherwise, refine shooting points $a_i^j \in \xi_i$ to ensure that the collinear condition holds true.
3. $j = j + 1$, **goto 2**.
4. **return** $\mathcal{Z} := \cup_{i=1}^k \mathcal{Z}_i^j$.

The Hausdorff distance between two sets A and B in 2D, denoted by $d_H(A, B)$, is defined by

$$d_H(A, B) = \max\{\sup_{x \in A} \inf_{y \in B} \|x - y\|, \sup_{y \in B} \inf_{x \in A} \|x - y\|\}.$$

By means of Hausdorff distance, the Algorithm 1 gives an appropriate shortest path between two points a and b in $\mathcal{D}$ by the following.

Theorem 3 ([1]). *Assume that a and b are points in a simple polygon $\mathcal{D}$ and $x_i \in SP(a, b), i = 0, 1, \ldots, k + 1, x_0 = a, x_{k+1} = b$. If $a_i \in \mathcal{D}$ such that $[a_i, x_i] \subset \mathcal{D}, 0 \le i \le k + 1$, then*

$$d_H\left(\cup_{i=1}^{k+1} SP(a_{i-1}, a_i), SP(a, b)\right) \le 2 \max_{0 \le i \le k+1} \|x_i - a_i\|.$$

2.4   Numerical Example

The algorithms are implemented in C++ code and run on Ubuntu Linux operating system with platform Intel Core i3, CPU 530 3.0HGz x 4. Fig. 4 shows a simple example of the shortest path between the vertex of smallest x-coordinate, denoted by a, and the vertex of largest x-coordinate, denoted by b, in the monotone polygon $\mathcal{PQ}$ of 40 vertices.

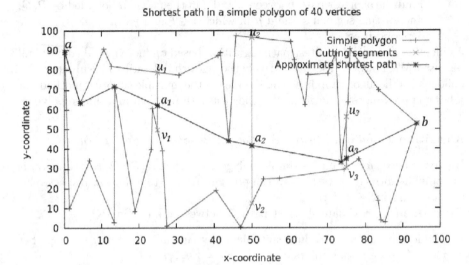

Fig. 4. The shortest path between a and b in $\mathcal{P}\mathcal{Q}$ is obtained by Algorithm 1, where $[u_i, v_i]$ is cutting segments, for $i = 1, 2, 3$. $SP(p, q) = \cup_{i=0}^{3} SP(a_i, a_{i+1})$, where $a_0 = a$, $a_4 = b$, and $a_i \in [u_i, v_i]$ for $i = 1, 2, 3$

3   Performance Analysis

3.1   Run Time

We now consider the run time of Algorithm 1. Due to the funnel algorithm a triangulation of polygon is first required. We used the implementation of the triangulation algorithm given in [9]. That is an improved version of the ear removal algorithm. The complexity is $O(n^2)$, where n is the number of vertices of the polygon. If we triangulate whole $\mathcal{D}$ then it will take a long time. By dividing $\mathcal{D}$ into sub-polygons $\mathcal{T}_i$ the size of polygons needed to triangulate, is then smaller. The time of triangulating sub-polygons is consequently reduced. Therefore, the total time of triangulating $\mathcal{T}_i$, for $i = 0, \ldots, k$, is theoretically small as a larger number of cutting segments.

The algorithm runs on the polygon $\mathcal{P}\mathcal{Q}$ in which $\mathcal{P}$ and $\mathcal{Q}$ are monotone with respect to x-coordinate without three collinear points. The number of vertices of the polygon is 350000. We run the algorithm for several sets of cutting segments. In Table 1, the run time of Algorithm 1 is reduced significantly as the cutting segments is increased. This was stated clearly in [1] as well. Here, we measure in more detail the time required for triangulating and dual tree constructing invoked in the funnel algorithm. Table 1 also shows that if the number of cutting segments is small then the run time of the triangulation and dual trees construction take almost the run time of the Algorithm 1. This indicates that if we choose an appropriate number of cutting segments, the method is efficient, specially in case of huge data.

Table 1. The run time (in seconds) of Algorithm 1 and the run times (in seconds) of triangulation and dual tree construction for the funnel algorithm

No. of cutting segments	Algorithm 1	Triangulation	Dual tree construction
0	41,828.75	21,129.27	16,325.46
5	7,344.34	4,464.89	2,723.89
10	4,060.89	2,493.57	1,467.05
50	858.65	497.55	192.49
100	504.99	251.51	96.27
500	284.62	51.08	19.05
1000	275.76	26.21	9.74

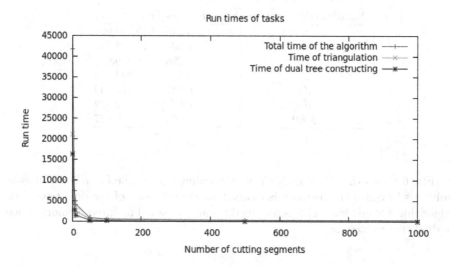

Fig. 5. The run time of Algorithm 1, triangulation, and dual tree constructing

However, the time of the refinement is also large comparing with the time of the triangulation and the dual tree construction as the number of cutting segments is increased. This means that the collinear condition of Algorithm 1 might be difficult to reach as a large number of cutting segments (see Fig. 5).

3.2 Memory Usage

Another task for the funnel algorithm is the dual tree construction. If each triangle of the triangulation is considered as a node of the tree, then the tree should be a binary one. An arbitrary triangle can be chosen to become the root of the tree. We start at the root and then construct the tree by adding new nodes based on the adjacent relationship between triangles. For each step of adding a new node, however, we need to traverse the current tree starting at the root. Therefore, a memory space of a node type must be allocated. This is why in case of huge data, the memory space required in the funnel algorithm is large. Again, the method given in [1] helps us to overcome this issue, because the triangulation is only required on sub-polygons.

In our implementation, tree traversal type is the in-order searching without recursion. Table 2 shows the peak memory usages in two cases. In the first case, if we free the memory cells of nodes after each searching, then the required memory is small. The second one requires more memory space, since the memory is only cleaned up after whole dual tree is constructed.

Table 2. The peak memory (MB) of Algorithm 1 in which the shortest paths in sub-polygons are computed by using the funnel algorithm

No. of cutting segments	Free on nodes	Free on trees
0	3.218	579,544.000
5	0.538	16,148.701
10	0.296	4,880.905
50	0.065	234.408
100	0.033	61.639
100	0.007	2.922
1000	0.004	0.827

Fig. 6 shows that if we implement Algorithm 1 using the funnel one for sub-polygons in which the memory is cleaned up at each node of the dual tree, then Algorithm 1 is efficient. This is obtained because of small required memory space of dual tree processing.

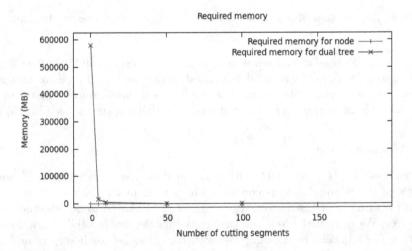

Fig. 6. The required memory is too large as number of cutting segments is reduced

4 Concluding Remarks

A triangulation-based multiple shooting method for solving the 2D geometric shortest path problems is implemented to show the performance of algorithm introduced in [1]. The run time of the polygon triangulation and the required memory of the dual tree construction are reduced significantly as number of cutting segments is increased. Numerical tests, however, also indicate that the number of cutting segments should be chosen appropriately. This follows the fact that if there are too many cutting segments, then the time of the refinement might be increased consequently.

Acknowledgments. This work was partially supported by the National Foundation for Science and Technology Development, Vietnam (NAFOSTED) under grant number 101.02-2011.45, the Portuguese National Funds through FCT (Fundação para a Ciência e Tecnologia) under the scope of project PEst-OE/MAT/UI0822/2011, and the Vietnam Institute for Advanced Study in Mathematics (VIASM).

References

1. An, P.T., Hai, N.N., Hoai, T.V.: Direct multiple shooting method for solving approximate shortest path problems. Journal of Computational and Applied Mathematics **244**, 67–76 (2103)
2. An, P.T., Hoai, T.V.: Incremental convex hull as an orientation to solving the shortest path problem. International Journal of Information and Electronics Engineering **2**(5), 652–655 (2012)
3. Aleksandrov, L., Maheshwari, A., Sack, J.-R.: Approximation algorithms for geometric shortest path problems, In: Proceedings of the 32nd ACM-STOC (Symposium on Theory of Computing), pp. 286–295, Portland, Oregon (2000)
4. Book, H.G., Plitt, K.J.: A multiple shooting method for direct solution optimal control problems. In: Proceedings of the 9th IFAC World Progress, pp. 225–236. Pergamon Press, Budapest (1984)
5. Chazelle, B., Guibas, L.: Visibility and intersection problems in plane geometry. In: Proceedings of 1st ACM Symposium on Computational Geometry, pp. 135–146 (1985)
6. Clarkson, K.: Approximation algorithms for shortest path motion planning. In: Proceedings of the 19th Annual ACM Symposium on Theory of Computing, pp. 56–65, New York (1987)
7. Demyen, D., Buro, M.: Efficient triangulation-based pathfinding. In: Proceedings of the 21st National Conference on Artificial Intelligence and the Eighteenth Innovative Applications of Artificial Intelligence Conference, pp. 942–947, Boston, Massachusetts, USA (2006)
8. Kallmann, M.: Path planning in triangulation. In: Proceedings of the Workshop on Reasoning. Representation, and Learning in Computer Games, International Joint Conference on Artificial Intelligence (IJCAI), pp. 49–54, Edinburgh, Scotland (2005)
9. O'Rourke, J.: Computational Geometry in C, 2nd edn. Cambridge University Press (1998)
10. Lee, T., Preparata, F.P.: Euclidean shortest paths in the presence of rectilinear barriers. Networks **14**, 393–410 (1984)

11. Mitchell, J.S.B.: Geometric shortest paths and network optimization. In: Sack, J.R., Urrutia, J. (eds.) Handbook of Computational Geometry, pp. 633–701. Elsevier Science B. V. (2000)
12. Sharir, M., Schorr, A.: On shortest paths in polyhedral spaces. SIAM Journal on Computing **15**(1), 193–215 (1986)
13. Stoer, J., Bulirsch, R.: Introduction to Numerical Analysis, 3rd edn. Springer, New York (2002)
14. Toussaint, G.T.: Computing geodesic properties inside a simple polygon. Revue D'Intelligence Artificielle **3**, 9–42 (1989)

Protecting Biometric Features by Periodic Function-Based Transformation and Fuzzy Vault

Thu Thi Bao Le$^{(\boxtimes)}$, Tran Khanh Dang, Quynh Chi Truong, and Thi Ai Thao Nguyen

Faculty of Computer Science and Engineering, HCMC University of Technology,
VNUHCM, Ho Chi Minh City, Vietnam
{thule,khanh,tqchi,thaonguyen}@cse.hcmut.edu.vn

Abstract. Biometrics-based authentication is playing an attractive and potential approach nowadays. However, the end-users do not feel comfortable to use it once the performance and security are not ensured. Fuzzy vault is one of the most popular methods for biometric template security. It binds a key with the biometric template and obtains the helper data. However, the main problem of fuzzy vault is that it is unable to guarantee the revocability property. In addition, most of the fuzzy vault schemes are performed on two biometrics modalities, fingerprints and iris. In previous works, authors suggested some cancelable transformations attached to a fuzzy vault scheme to overcome these weaknesses. However, the computational cost of these proposals was quite large. In this paper, we present a new hybrid scheme of fuzzy vault and periodic function-based feature transformation for biometric template protection. Our transformation is not only simpler but also suitable for many kinds of biometrics modalities. The newly proposed fuzzy vault scheme guarantees the revocability property with an acceptable error rate.

Keywords: Biometrics · Fuzzy vault · Biometrics template protection · Cancelable transformations · Face recognition

1   Introduction

As we all know, the traditional authentication schemes usually based on something the user has (such as: smart card) or something the user knows (such as: password, PIN). However, those techniques have several limitations. For example, they cannot distinguish between an authorized user and those who know the correct password [1]. So, we have to choose a strong password and always to keep it in mind.

In recent years, with the rapid development of technologies, biometrics-based authentication systems are becoming potential, because biometrics is literally stuck to an individual, it can prevent the use of several identities by a single individual. The term biometric (from the Greek for bio=life, metric=degree) refers to authentication by means of biological (more accurately, physiological or behavioral) features (such as: face, voice, fingerprint...) [2]. Using biometrics can overcome above limitations. But, it still raises some security and privacy concerns [1]. For example, biometrics is secure but not secret, because voice, face... can be easily recorded and may be misused

© Springer-Verlag Berlin Heidelberg 2014
A. Hameurlain et al. (Eds.): TLDKS XVI, LNCS 8960, pp. 57–70, 2014.
DOI: 10.1007/978-3-662-45947-8_5

without the user's consent.   Another problem is that unlike passwords, cryptographic keys, or PINs, biometrics cannot be changed once compromised. In addition, a user can be tracked by means of cross-matching when he/she uses the same biometrics across all applications and the service-providers collude with each other.

Therefore, the security of biometric template has been emerging increasingly and a lot of research has been done in this field. According to the authors in [3], there are four properties that an ideal biometric template protection scheme should possess:

1. Diversity: the secure template must not be the same in two different applications; therefore, the user's privacy is ensured.
2. Revocability: it should be straightforward to revoke a compromised template and reissue a new one based on the same biometric data.
3. Security: An original biometric template must be computationally hard to recover from the secure template. This property guarantees that an adversary does not have the ability to create a physical spoof of the biometric trait from a stolen template.
4. Performance: the biometric template protection scheme should not degrade the recognition performance of the biometric system.

In biometric template protection, fuzzy vault is considered as a popular method. It binds a key with the biometric template and obtains the helper data for authentication. The template is hidden in the helper data. However, there are some problems that fuzzy vault encounters with. One of these stems from the reason that fuzzy vault cannot provide the diversity and revocability properties. In this paper, that shortcoming is made good by applying a feature transformation in a fuzzy vault scheme. Strictly speaking, the main idea for the marriage of fuzzy vault with feature transformation was introduced in a few recent proposals. Nevertheless, majority of which focus on two biometrics modalities, fingerprints [4, 5, 6], and iris [7, 8]. The transformations for face based fuzzy vault scheme are rare and very complicated. Therefore, this paper will present a hybrid scheme of face based fuzzy vault and feature transformation. Our proposed transformation is not only simpler but also suitable for many kinds of biometrics modalities. The face biometric templates are protected by hidden in the set of chaff points generated by fuzzy vault scheme. Besides, these templates are able to be changed or revoked if the owners have suspicious about being tracked or stolen. Our experimental result will show that the newly proposed scheme guarantees the revocability property with an acceptable error rate.

The structure of this paper is organized as follows. Section 2 provides a brief review of related works. The details of our proposed scheme are described in Section 3. Following them, the evaluation is discussed in Section 4. At last, Section 5 provides the conclusion and future works.

2   Related Works

Biometric template protection is an important issue in a biometric system. Biometric template of a person cannot be replaced or used again once it is compromised. In [9], the authors presented two approaches to deal with this issue, including feature transformation and biometric cryptosystem.

In the biometric cryptosystem approach, a key is derived from the biometric template or bound with the biometric template. Both the biometric template and the key are then discarded, and only the public helper data is stored in the database. Although public helper data does not reveal any information about the biometrics and the key, it is very useful to regenerate the key from another biometric sample which is closed to the biometric template. The concepts of secure sketch and fuzzy extractor [10], a combination of ANN and secure sketch [11] are kinds of biometric cryptosystem approach. The fuzzy commitment scheme [12] and fuzzy vault [13] are two examples of the key binding approach.

In the feature transformation approach, the biometric templates are transformed before being stored in the database. The transformed templates are hard to be recovered to the original template even with some knowledge of transformation function. Then, the transformed templates are safe to store in the database.

2.1 Fuzzy Vault

Juels and Sudan [13] introduced a construct called a fuzzy vault. The idea is that Alice places a secret k in a fuzzy vault and locks it using a set A of elements from some public universe U. To unlock the vault and retrieve k, Bob must present a set B closed to A, i.e., B and A overlap substantially.

To construct a fuzzy vault, first, Alice selects a polynomial p of variable x that encodes k. Considering the elements of A as distinct x-coordinate values, she computes the polynomial projections for the elements of A. Then, she adds some randomly generated chaff points that do not lie on p. The final set includes real points which lie on p and chaff points. The number of chaff points is far greater the number of real points. It will make the attacker hard to find the real points.

When Bob want to unlock the vault and learn k (i.e., find p), he uses his unordered set B. If B overlaps with A substantially, he will be able to locate many points in the vault that lie on p. By using error-correction coding (e.g., Reed-Solomon), it is assumed that he can reconstruct p and discover k.

There are many researches follow this scheme to construct the vault for fingerprint [4, 5, 6], iris [7, 14], face [15], and some other biometric types.

However, several attacks against fuzzy vaults have been discovered [16, 17]. These are: attacks via record multiplicity, stolen key inversion attack and blended substitution attack. In a stolen key inversion attack, if an adversary somehow recovers the key embedded in the vault, he can decode the vault to obtain the biometric template. Because the vault contains a large number of chaff points, it is possible for an adversary to substitute a few points in the vault with his own biometric features. In this case, the system allows both the genuine user and the adversary to be successfully authenticated. This attack is called blended substitution. In record multiplicity attack, an adversary can access to two different vaults generated from the same biometric data (from two different applications). He can easily identify the genuine points in the two vaults and decode the vault. Thus, the fuzzy vault scheme does not provide diversity and revocability properties. In this paper, we proposed a hybrid scheme where biometric templates are first transformed based on a periodic function to guarantee diversity and revocability properties.

2.2   Feature Transformation

Transformation functions are classified into two types: invertible (or salting) and non-invertible transformation.

Salting is a method in which the biometric features are transformed using a function defined by a user-specific key or password [9]. With the key, we can invert the transform template to the original one. Therefore, the key needs must be kept secret. Salting can be considered as two-factor authentication in which the users must present both secret key and biometric trail to the authentication system. In [18], the authors generate a user-based random orthonormal $n * n$ matrix A, where n is the size of biometric feature vectors. Then, the original template feature vector x is transformed to a secure domain using matrix product: y = Ax. The random orthonormal matrix is generated from a user-based key or token using Gram-Schmidt algorithm[1]. The security in this scheme is relied on the user-specific random matrix which plays a role as a secret key. Another example of salting is using a user-based shuffling key to transform an iris code in [19]. User-based shuffling key which is generated based on users' key or password is an n-bit string. An iris code is also divided into n blocks. The transformation works as follows: beginning from the first to the last block, if the bit i^{th} is 1 (or 0), block i^{th} will be moved to the first (or last) place of the code.

In non-invertible transformation, the biometric template is transformed by a one-way transformation function. A one-way function F is "easy to compute" (in polynomial time) but "hard to invert" [9]. The function F can be public. Non-invertible transformation for fingerprint is proposed in [20]. The authors presented three methods to transform fingerprint. In the first method, the fingerprint image is divided into rectangular grid cells. A shifting map is defined as a transformation function. The minutiae in each cell are moved to a new position which is defined in a shifting map. There may be some minutiae to be shifted to the same cell. Thus, even if the shifting map is public, the attacker cannot infer that a minutia in the transformed template is belonged to which cell in the original template. This is the characteristic of non-invertible transformation. Similarly, in the second method, the fingerprint image is divided into sectors, and the minutiae are shifted among sectors which have the same or nearly the same radius. The third method considers not only the position but also the direction of the minutiae. Scutu et al. [21] proposed a secure authentication based on robust hashing. The idea is to embed each component of a feature vector into a Gaussian function. After that, a number of fake Gaussians are added to hire the true Gaussian.

To all of noninvertible transformation, the most challenge is that how to preserve the similarity of distances among transformed templates and among original tem-

---

[1]   Gram–Schmidt algorithm from Wikipedia: http://en.wikipedia.org/wiki/Gram%E2%80% 93Schmidt_process (Oct 2014).

plates. It means that two transformed templates must be closed if the two original templates are closed. This characteristic keeps the error rates of the transformed biometric systems similar to the generic biometric systems, but the transformed biometric systems protect the templates from being compromised.

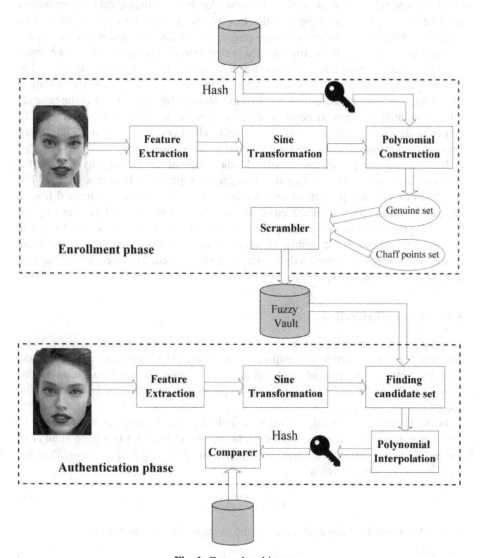

Fig. 1. General architecture

3   The Proposed Scheme

3.1   General Architecture

Our proposed scheme can be applied for many kinds of biometric data whose feature is in vectors. In this paper, we use the face biometric data for demonstration. Our general architecture includes two main phase: enrollment and authentication. In enrollment phase, a set of biometric features is first extracted from the users' face images. After standardized, these features are then transformed by a sine function. The randomly generated key is used to construct a polynomial function and its hashing is stored for matching purpose in authentication phase. The transformed features apply this polynomial function to generate a set of genuine points in fuzzy vault set. To complete the fuzzy vault encoding step, a set of chaff points is also added into fuzzy vault set. After that, all these points are stored in fuzzy vault database.

In authentication phase, sensor will take the image of a user and provide it for the system. This image is also extracted in order to gain the user's biometric feature. The sine transformation is performed on the extracted feature. If this transformed feature has substantial overlap with the enrolled ones, the secret key will be correctly retrieved by the fuzzy vault decoding step. Afterwards, the recovered key is hashed in order to compare with the hashed versions of the keys generated in the enrollment phase. If the new key is matched, the user is authenticated. The overview of the system is illustrated in the Fig. 1.

3.2   Feature Extraction

A feature extraction technique is used to extract the biometric feature. Among many different feature extraction techniques, PCA (Principal Component Analysis), and ICA (Independent Component Analysis) are popular ones for face recognition. In this paper, we choose PCA for its significant outperformance on human face recognition task [22].

In the Eigenfaces method, the PCA is applied to the training set to find a set of standardized face ingredients, called eigenfaces. The training set is a large number of images depicting different human faces, including $\Gamma_1, \Gamma_2, \Gamma_3, \dots, \Gamma_M$ images. We defined the average face of set as:

$$\psi = \frac{1}{M}\sum_{i=1}^{M} \Gamma_i \tag{1}$$

The difference between each face and the average is shown by vector:

$$\phi_i = \Gamma_i - \psi \tag{2}$$

Then, the covariance matrix is calculated by:

$$C = \frac{1}{M}\sum_{i=1}^{M} \phi_i \phi_i^T = AA^T \tag{3}$$

where the matrix $A = [\phi_1 \phi_2 \dots \phi_M]$.

We can obtain M eigenvector and eigenvalue of covariance matrix C. Then, we sort out R (R ≤ M) largest eigenvectors by the corresponding eigenvalues, denoted as: $U = [u_1 u_2 \ldots u_R]_{N^2 xR}$. u_i is the eigenface, these eigenfaces are orthogonal to each other. The image of a user can be transformed to the R-dimensional face space by linear mapping:

$$\Omega = U^T (\Gamma - \psi) = \begin{bmatrix} \omega_1 \\ \omega_2 \\ \vdots \\ \omega_R \end{bmatrix}_{Rx1} \tag{4}$$

3.3   Sine Transformation

Before going into details about our transformation, we introduce some notations which are often used in this section.

− $X = \{x_1, x_2, \ldots, x_n\}$: a biometric feature vector extracted from the face image of a user.
− $Y = \{y_1, y_2, \ldots, y_n\}$: an intermediate vector after applied sine transformation on X.
− $X' = \{x'_1, x'_2, \ldots, x'_n\}$: a final transformed vector which is used for fuzzy vault encoding and decoding.

When the feature extraction step completes, the feature vector X of all users will be transformed into an intermediate vector Y by the function (5):

$$\sin (x_i + y_i) = c \ (c \text{ is chosen randomly}) \tag{5}$$

As we all know, sine function has period of 2 Π rad. So, with each value of x_i, we will find exactly one value y_i. But, given a value of y_i, you cannot derive an exactly x_i, because there are many value of x_i corresponding with that y_i. In another words, the sine function is a non-invertible transformation. You can also choose another periodic function has the same characteristic with sine function (such as cosine function). In this paper, we use the sine function to present our works.

The value of y_i is obtained by finding the minimum $y_i > 0$ such as:

$$y_i = \sin^{-1}(c) - x_i \tag{6}$$

It means that the value of y_i lies between [0; 2Π]. The sine transformation is simply illustrated in the Fig. 2. For example, assume that c = 1, x_i = Π/3, so we have y_i = 2 Π/3.

However, the fuzzy vault requires that the points in vault are disordered. So we cannot use this y_i as x-coordinate values, because if so, it will eliminate the ordered property of feature vector Y. This reason causes the increase of FAR (False Accept Rate). To eliminate this risk, we apply a minor transformation to the vector Y for preserving the order of the elements in fuzzy vault set. The result is the final transformed vector X' generated by the following rules:

$$x'_1 = y_1 \tag{7}$$

$$x'_2 = \begin{cases} y_2 \text{ if } y_2 < x'_1 \\ y_2 + k\pi, otherwise \ (where \ k \ is \ minimum \ such \ as \ y_2 + k\pi > x'_1) \end{cases} \tag{8}$$

$$\dots.$$

$$x'_n = \begin{cases} y_n \text{ if } y_n < x'_{n-1} \\ y_n + k\pi, otherwise \ (where \ k \ is \ minimum \ such \ as \ y_n + k\pi > x'_{n-1} \end{cases} \tag{9}$$

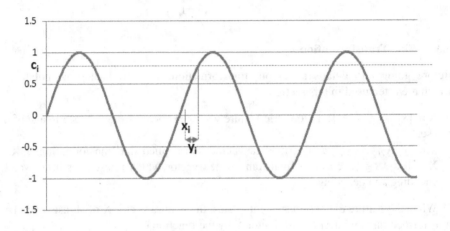

Fig. 2. Sine transformation

3.4   Improve the Performance of Sine Transformation

Easy to see that, the minor transformation (7), (8), (9) in section 3.3 can increase error rate of the system. We have to suppose another way to satisfy the disordered feature of fuzzy vault. The new transformation is:

$$\sin (x_i + y_i) = c_i \ (c_i \text{ is chosen randomly between } [-1, 1]) \tag{10}$$

In addition, the sine function can map two points with large distance to new two closed points. One example is shown in Fig. 3.

To avoid this phenomenon, the value of y_i is obtained by:

$$y_i = \sin^{-1}(c_i) - x_i, \text{ where } \sin^{-1}(c_i) = t + k2\Pi \text{ and } t = [\frac{-\Pi}{2}, \frac{\Pi}{2}] \tag{11}$$

It means that, we eliminate another solution: $\sin^{-1}(c_i) = t + k2\Pi$ and $t = [\frac{\Pi}{2}, \frac{3\Pi}{2}]$. Note that y_i can be a negative number. The value of y_i lies between $[-2\Pi, 2\Pi]$. This new transformation is illustrated in Fig. 4. The output value y_i will be used as an input to fuzzy vault scheme presenting in next sections.

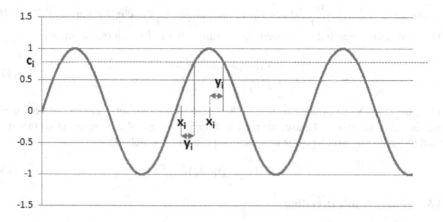

Fig. 3. Mapping two distant points into new two closed points

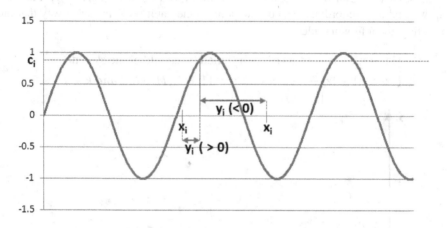

Fig. 4. New sine transformation

3.5   Fuzzy Vault Encoding

The key K is 144-bit which is randomly generated by a random number generator. This key is used for polynomial construction in fuzzy vault scheme. At first, a 8-order polynomial, $f(x) = c_8 x^8 + c_7 x^7 + \cdots + c_1 x + c_0$, needs to be generated. The values of these coefficients are created by truncating the 144-bit K to 9 non-overlapping 16-bit segments. And then, each of them is mapped to the coefficients $c_8 - c_0$ in succession. For example, the first 16 bits is mapped to c_8, and so on. The order of the mapping should be preserved for encoding and decoding of the vault.

To create the genuine points, the polynomial $f(x)$ is evaluated on each of the transformed feature points x'_i. As a result, the genuine set G consists of a set of pairs $\{x'_i, f(x'_i)\}_{i=1}^M$, where M is the dimension of the feature vector (also is the dimension of the transformed features). The next step we need to do is to generate the chaff points set $C = \{a_j, b_j\}_{j=1}^{N_C}$, where N_C is the number of the chaff points and $N_C \gg M$. The randomly generated set C needs to guarantee the following requirements:

$$\begin{cases} |a_j - x'_i| > \Delta & \forall i \ (\Delta \neq 0) \\ b_j \neq f(a_j) & \forall j \end{cases} \tag{12}$$

The final vault V is obtained by taking the union of the two sets G and C. Before storing the set V into the database, we pass it through a scrambler component so that it is hard to figure out which points are genuine points and which are not.

$$V = C \cup G = \{r_k, s_k\}_{k=1}^{M+N_C} \tag{13}$$

3.6   Fuzzy Vault Decoding

The fuzzy vault decoding is mainly based on the Lagrange polynomial interpolation. Assume that the transformed feature vector of a user is $Z = \{z_1, z_2, ..., z_M\}$. For each z_i, we find the x-coordinate (r_j) of one point in the vault set V in such a way that the r_j satisfy the following rules:

$$\begin{cases} |z_i - r_j| \leq \varepsilon, \ \varepsilon \ll \Delta \ (\varepsilon \text{ is a designed threshold of the system}) \\ |z_i - r_j| \leq |z_i - r_k|, \qquad k = \{1, 2, ..., M + N_C\} \ and \ k \neq j \end{cases} \tag{14}$$

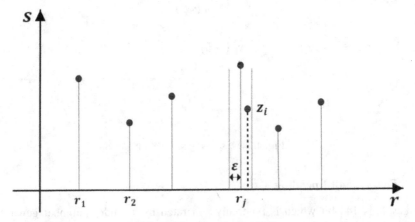

Fig. 5. Fuzzy vault decoding

As a result, the set of points $\{r_j, s_j\}_{j=1}^{L \leq M}$, whose r_j has just found, is the set of the candidate points. These points are ranked by the corresponding nearest distance between r_j and z_i. To recover the 8-order polynomial, the Lagrange interpolation technique [2] needs 9 points. We choose the first I points ($9 \leq I$) of the ranked candidate set (the points have the highest possibility to be the real points) and then make the

---

[2] Wolfram MathWorld, Lagrange Interpolating Polynomial: http://mathworld.wolfram.com/ LagrangeInterpolatingPolynomial.html (Oct 2014).

combinations of 9 points from the I ranked candidate set (C_I^9). For each combination, we find one polynomial. Its coefficients are mapped back and concatenated in the same order as encoding phase in order to obtain a 144-bit key K'. To check whether the key K' is matched with the initial K or not, we hash K' and compare the result with the hashed versions of the keys in the database. The authentication is successful if and only if we can recover one key K' matched with K. It means that we do not have to compute all the combinations. Otherwise, if no matched key from all the combinations is found, the authentication is failed.

4 Evaluation

The proposed scheme with the original transformation (section 3.3) and new transformation (section 3.4) is tested with the Face94 database[3]. In the training procedure, we use 100 images of 50 people, i.e. 2 images per person, to construct 40 eigenfaces. Then, we test the scheme with 152 people, including 50 people participated in the training process and 102 new people. Each person has 5 images in which 1 is used to create the vault and 4 are used to unlock the vault. We measure the performance of this scheme, include: FAR (False Acceptance Rate) and FRR (False Rejection Rate)[4].

— The FAR defines the probability that the system incorrectly matches the input pattern to a non-matching template in the database. It measures the percent of invalid inputs which are incorrectly accepted.
— By analogy, the FRR defines the probability that the system fails to detect a match between the input pattern and a matching template in the database. It measures the percent of valid inputs which are incorrectly rejected.

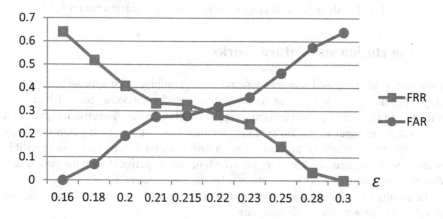

Fig. 6. FAR and FRR of proposed scheme with original transformation

[3] Libor Spacek's Faces94 database: http://cswww.essex.ac.uk/mv/allfaces/faces94.html. (Oct 2014).
[4] FAR and FRR from Wikipedia: http://en.wikipedia.org/wiki/Biometrics (Oct 2014).

The results of are shown in the Fig. 6 and Fig. 7 correspond to original transformation and new transformation. The vertical axis is the error rate which is range [0, 1], and the horizontal axis is the epsilon (ε) as defined in Section 3.6. Epsilon is the minimum distance among points in the vault. Using the original transformation, we can get the acceptable error rates at $\varepsilon = 0.22$. With this value, the error rates are: $FRR \approx FAR \approx 0.3$. Otherwise, using the new transformation, we can get the better error rates $FRR \approx FAR \approx 0.25$ at $\varepsilon = 0.0175$.

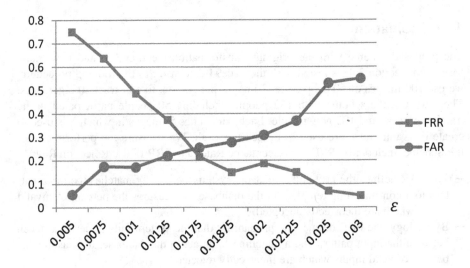

Fig. 7. FAR and FRR of proposed scheme with new transformation

5   Conclusion and Future Works

In this paper, we proposed a hybrid scheme which combines two approaches, namely fuzzy vault and periodic function-based feature transformation, to protect biometric templates. In this scheme, we perform a transformation on the biometric template and then let it as an input to the fuzzy vault. By this way, we can strengthen the fuzzy vault with the revocability property. Our transformation function is non-invertible because the transformation function, i.e. sine function, is periodic with the period 2π. With the knowledge of sin(x), we cannot infer the true value of x.

The results of the evaluation confirm the effective and the practical properties of our scheme to protect biometric template.

The next works we do with our research are to reduce the error rates (FAR, FRR) so that it can be used in practice and to find a proper way to add chaff points to the vault. For the first task, we need more researches on the range of each component of the feature vectors so that we can adjust our parameters, namely the minimum distance among points in the vault and the maximum range of x in the vault. If we can decide these parameters more precisely we can archive a lower error rates in our scheme. The other approach is that we can apply multimodel biometrics schema to improve performance of the system [23]. For the second task, in this paper, the chaff points are added randomly without the consideration that the transform values (with respect to x value in the vault) of true points have a maximum distance of a predefined number of pi to their nearest true points. This property can be exploited to limit an amount of possible cases in a brute force attack. Therefore, we need to find a better solution to add chaff points to the vault in our scheme.

Acknowledgements. This research is funded by Vietnam National University - Ho Chi Minh City (VNU-HCM) under grant number B2013-20-02. We also want to show a great appreciation to each member of D-STAR Lab (www.dstar.edu.vn) for their enthusiastic supports and helpful advices during the time we have carried out this research.

References

1. Ratha, N., Chikkerur, S., Connell, J., Bolle, R.: Privacy Enhancements for Inexact Biometric Templates. In: Security with Noisy Data, pp. 153–168. Springer, London (2007)
2. Salomon, D.: Elements of Computer Security. Springer (2010). 978-0-85729-005-2
3. Maio, D., Jain, A.K.: Handbook of fingerprint recognition. Springer (2009)
4. Clancy, T C., Kiyavash, N., Lin, D.J.: Secure smartcard-based fingerprint authentication. In: Proceedings of the 2003 ACM SIGMM Workshop on Biometrics Methods and Applications, pp. 45–52. ACM (2003)
5. Nandakumar, K., Jain, A.K., Pankanti, S.: Fingerprint-based fuzzy vault: Implementation and performance. IEEE Transactions on Information Forensics and Security 2(4), 744–757 (2007)
6. Uludag, U., Jain, A.K.: Fuzzy fingerprint vault. In: Proceedings of Workshop Biometrics: Challenges Arising from Theory to Practice, pp. 13–16 (2004)
7. Lee, Y.-J., Bae, K., Lee, S.-J., Park, K.R., Kim, J.H.: Biometric key binding: Fuzzy vault based on iris images. In: Lee, S.-W., Li, S.Z. (eds.) ICB 2007. LNCS, vol. 4642, pp. 800–808. Springer, Heidelberg (2007)
8. Hao, F., Anderson, R., Daugman, J.: Combining crypto with biometrics effectively. IEEE Transactions on Computers 55(9), 1081–1088 (2006)
9. Jain, A.K., Nandakumar, K, Nagar, A.: Biometric template security. EURASIP Journal on Advances in Signal Processing 2008(113) (2008)
10. Dodis, Y., Reyzin, L., Smith, A.: Fuzzy extractors: How to generate strong keys from biometrics and other noisy data. In: Cachin, C., Camenisch, J.L. (eds.) EUROCRYPT 2004. LNCS, vol. 3027, pp. 523–540. Springer, Heidelberg (2004)
11. Huỳnh, V.Q.P., Thai, T.T.T., Dang, T.K., Wagner, R.: A Combination of ANN and Secure Sketch for Generating Strong Biometric Key. Journal of Science and Technology, Vietnamese Academy of Science and Technology 51(4B), 203–212 (2013). ISSN 0866-708X
12. Juels, A., Wattenberg, M.: A fuzzy commitment scheme. In: Proceedings of the 6th ACM Conference on Computer and Communications Security, pp. 28–36. ACM (1999)
13. Juels, A., Sudan, M.: A fuzzy vault scheme. Designs, Codes and Cryptography 38(2), 237–257 (2006)
14. Vo, T.T.L., Dang, T.K., Küng, J.: A Hash-Index Method for Securing Fuzzy Vaults. In: Eckert, C., Katsikas, Sokratis K., Pernul, G. (eds.) TrustBus 2014. LNCS, vol. 8647, pp. 60–71. Springer, Heidelberg (2014)

15. Wu, Y., Qiu, B.: Transforming a pattern identifier into biometric key generators. In: IEEE International Conference on Multimedia and Expo (ICME), pp. 78–82 (2010)
16. Scheirer, W.J., Boult, T.E.: Cracking fuzzy vaults and biometric encryption. In: Proceedings of the Biometrics Symposium, Baltimore, Md, USA (September 2007)
17. Nguyen, M.T., Truong, Q.H., Dang, T.K.: Enhance Fuzzy Vault Security using Nonrandom Chaff Point Generator. In: Dang, T.K., Wagner, R., Neuhold, E., Takizawa, M., Küng, J., Thoai, N. (eds.) FDSE 2014. LNCS, vol. 8860, pp. 204–219. Springer, Heidelberg (2014)
18. Jin, A.T.B., Ling, D.N.C., Goh, A.: Bio hashing: two factor authentication featuring fingerprint data and tokenised random number. Pattern Recognition **37**(11), 2245–2255 (2004)
19. Kanade, S., et al.: Three factor scheme for biometric-based cryptographic key regeneration using iris. In: Biometrics Symposium 2008 (BSYM 2008), pp. 59–64. IEEE (2008)
20. Ratha, N.K., Chikkerur, S., Connell, J.H., Bolle, R.M.: Generating cancelable fingerprint templates. IEEE Transactions on Pattern Analysis and Machine Intelligence **29**(4), 561–572 (2007)
21. Sutcu, Y., Sencar, H.T., Memon, N.: A secure biometric authentication scheme based on robust hashing. In: Proceedings of the 7th Workshop on Multimedia and Security, pp. 111–116. ACM (2005)
22. Baek, K., Draper, B.A., Beveridge, J.R., She, K.: PCA vs. ICA: A Comparison on the FERET Data Set. In: Proc. of the 4th International Conference on Computer Vision (ICCV 2002), pp. 824–827 (2002)
23. Nguyen, V.N., Nguyen, V.Q., Nguyen, M.N.B., Dang, T.K.: Fuzzy Logic Weight Estimation in Biometric-Enabled Co-authentication System. In: Linawati, Mahendra, M.S., Neuhold, E.J., Tjoa, A.M., You, I. (eds.) CT-EurAsia 2014. LNCS, vol. 8407, pp. 365–374. Springer, Heidelberg (2014)

EPOBF: Energy Efficient Allocation of Virtual Machines in High Performance Computing Cloud

Nguyen Quang-Hung[✉], Nam Thoai, and Nguyen Thanh Son

Faculty of CSE, HCMC University of Technology, VNUHCM,
268 Ly Thuong Kiet Street, Ho Chi Minh City, Vietnam
{hungnq2,nam,sonsys}@cse.hcmut.edu.vn

Abstract. Cloud computing has become more popular in provision of computing resources under virtual machine (VM) abstraction for high performance computing (HPC) users. A HPC cloud is such a cloud computing environment. One of the challenges of energy-efficient resource allocation of VMs in HPC clouds is the trade-off between minimizing total energy consumption of physical machines (PMs) and satisfying Quality of Service (e.g. performance). On the one hand, cloud providers want to maximize their profit by reducing the power cost (e.g. using the smallest number of running PMs). On the other hand, cloud customers (users) want highest performance for their applications. In this paper, we study energy-efficient allocation of VMs that focuses on scenarios where users request short-term resources at fixed start-times and non-interrupted durations. We then propose a new allocation heuristic (namely Energy-aware and Performance-per-watt oriented Best-fit (EPOBF)) that uses performance-per-watt as a metric to choose which most energy-efficient PM for mapping each VM (e.g. the maximum of MIPS/Watt). Using information from Feitelsons Parallel Workload Archive to model HPC jobs, we compare the proposed EPOBF to state-of-the-art heuristics on heterogeneous PMs (each PM has multicore CPUs). Simulations show that the proposed EPOBF can significantly reduce total energy consumption when compared with state-of-the-art allocation heuristics.

Keywords: Cloud computing · VM allocation · Energy efficiency

1 Introduction

Cloud computing has been developed as a utility computing model [8] and is driven by economies of scale. High Performance Computing (HPC) clouds have been popularly adopted [15,18,20,23] and are provided by industrial companies such as Amazon Web Service Cloud [1]. A HPC cloud is a cloud system that uses virtualization technology for provision of computational resources in form of virtual machines (VMs) to run their HPC applications [11,18]. These cloud systems are often built from virtualized data centers [5,22]. Powering these cloud

© Springer-Verlag Berlin Heidelberg 2014
A. Hameurlain et al. (Eds.): TLDKS XVI, LNCS 8960, pp. 71–86, 2014.
DOI: 10.1007/978-3-662-45947-8_6

systems is very costly and is increasing with the increasing scale of these systems. Therefore, advanced scheduling techniques for reducing energy consumption of these cloud systems are highly concerned for any cloud providers.

Energy-efficient scheduling of HPC jobs in HPC cloud is still challenging [11,15,24,25]. One of the challenges of energy-efficient scheduling algorithms is the trade-off between minimizing energy consumption and satisfying Quality of Service (e.g. performance or resource availability on time for any reservation request [22,23]) [5]. Resource requirements are application-dependent. However, HPC applications are mostly CPU-intensive and, as a result, they could be unsuitable for dynamic consolidation and migration techniques as shown in [5,6] on HPC jobs/applications to reduce energy consumption of physical machines.

The Green500 list [21], which has been presented since 2006, has become popular. The idea of the Green500 list ranks HPC systems based on a metric of performance-per-watt (FLOPS/Watt), implying that the higher FLOPS/Watt, the more energy-efficient HPC system. Inspired by the idea of the Green500 list [21], in this paper, we propose a new VM allocation algorithm (called EPOBF) with two heuristics EPOBF-ST and EPOBF-FT which sort the list of VMs by starting time and respectively by finishing time. EPOBF uses similar metric of performance-per-watt to choose the most energy-efficient PM for mapping each VM. We propose two methods to calculate the performance-per-watt values. We have implemented the EPOBF-ST and EPOBF-FT heuristics as an extra VM allocation heuristic in the CloudSim version 3.0 [9]. We compare the proposed EPOBF-ST and EPOBF-FT heuristics to popular VM allocation heuristics which are PABFD (Power-Aware Best-fit Decreasing) [6], and vector bin packing greedy L1/L2/L30 (VBP Greedy L1/L2/L30) heuristics. The PABFD [6] is a best-fit heuristic to choose which PM has least increasing power on placement of each VM. The VBP Greedy L2/L1/L30 is a vector bin-packing norm-based greedy L2/L1/L30 in [19]. We evaluate these heuristics by simulations with a large-scale simulated system model, which has 10,000 heterogeneous PMs and simulated workload with thousands of cloudlets where each cloudlet can model a HPC task. These simulated cloudlets use information that is converted from a Feitelsons Parallel Workload Archive [2] (SDSC-BLUE-2000-4.1-cln.swf [3]) to model simulated HPC workload. Simulations show that both versions of EPOBF-ST and EPOBF-FT can reduce the total energy consumption by 21% and 35% respectively on average when compared with the PABFD, and both of versions of EPOBF-ST and EPOBF-FT can reduce 38% and 49% respectively in comparison to the vector bin-packing norm-based greedy heuristic.

The rest of this paper is structured as follows. Section 2 discusses related works. Section 3.2 introduces the system model that includes energy-aware HPC cloud architecture, power model and our proposed EPOBF-ST and EPOBF-FT that is a power-aware scheduling algorithm using two ways to rank physical machines by performance-per-watts. Section 5 discusses simulated experiments on the algorithms: EPOBF-ST and EPOBF-FT, PABFD (based line), and vector bin-packing norm-based greedy (VBP Greedy L1, VBP Greedy L2, and VBP Greedy L30). Section 6 concludes this paper and introduces future works.

2   Related Works

Cloud computing has been developed as an utility computing model [8] and is driven by economies of scale. Sotomayor et al. [22,23] proposed a lease-based model and implemented First-Come-First-Serve (FCFS) scheduling algorithm and a greedy-based VM mapping algorithm to map leases that include some of VMs with/without start time and user specified duration to a set of homogeneous physical machines (PMs). To maximize performance, these scheduling algorithms tend to choose free load servers (i.e. those with the highest-ranking scores) when allocating new VMs. On the other hand, the greedy algorithm can allocate a small lease (e.g. with one VM) to a multicore physical machine. As a result, the greedy algorithm cannot optimize for energy efficiency.

Many works have considered the VM placement problem as a bin-packing problem. They use bin-packing heuristics (e.g. First-Fit Decreasing (FFD) and Best-Fit Decreasing (BFD)) to place virtual machines (VMs) onto a minimum number of physical servers to minimize energy consumption [5,6]. Microsoft research group [19] has studied first-fit decreasing (FFD) based heuristics for vector bin-packing to minimize number of physical servers in the VM allocation problem. Beloglazov et al. [5,6] have proposed VM allocation problem as bin-packing problem and presented a power-aware modified best-fit decreasing (denoted as PABFD) heuristic. PABFD sorts all VMs in a decreasing order of CPU utilization and tends to allocate a VM onto an active physical server that would take the minimum increase of power consumption. However, choosing a host with a minimized increasing power consumption does not necessarily imply minimizing total energy consumption in VM allocation problems where all physical servers are identical and the power consumption of a physical server is linear to its CPU utilization. The PABFD prefers to allocate a VM to a host that will increase least power consumption. On the other hand, the PABFD can assign VMs to a host that has a few cores and the authors are only concerned about CPU utilization. The PABFD also does not consider the starting time and finishing time of these VMs. Therefore, it is unsuitable for the power-aware VM allocation considered in this paper, i.e. the PABFD cannot result in a minimized total energy consumption for VM placement problem with certain interval time while still fulfilling the quality-of-service (e.g. performance or resource availability on time for any reservation request [22,23]).

Goiri et. al. [12] has developed a score-based scheduling which is a hill-climbing algorithm searching for best match (host,VM) pairs. In their work, score of each (host,VM) pair is the sum of many factors such as power consumption, hardware and software fulfillment, resource requirement. In contrast, our proposed EPOBF chooses a host that has a maximum of MIPS/Watts to assign a VM. We are concerned about three resource types: processing power (e.g. MIPS), size of physical memory and network bandwidth and energy consumption. Previous studies such as [5,6,12,19] are suitable for service allocation, in which each VM will execute a long running, persistent application. In contrast, our proposed EPOBF considers the case where each user VM has a certain interval time (i.e. started at a starting time in non-preemptive duration). We consider provision of resources for HPC applications that will start at a fixed point in time for a non-interrupted duration. This makes our paper distinguished from the previous works that survey in [7,13,16].

Some other research [11,15] considers HPC applications/jobs in HPC cloud. Garg et al. [11] proposed a meta-scheduling problem to distribute HPC applications to cloud systems with distributed N data centers. The objective of scheduling is minimizing CO_2 emission and maximizing the revenue of cloud providers. Le et al. [15] distribute VMs across distributed cloud virtualized data centers whose electricity prices are different in order to reduce the total electricity cost.

Research on energy-efficient job scheduling algorithms use Dynamic Voltage Frequency Scaling (DVFS)-based mechanism is active. Albers et al. [4] reviewed some energy efficient algorithms which were used to minimize flow time by changing processor speed adapted to job size. Some works [14,24] proposed scheduling algorithms to flexibly change processor speed in such a way that meets user requirements and reduces power consumption of processors when executing user applications. Laszewski et al. [14] proposed scheduling heuristics and presented application experiences for reducing power consumption of parallel tasks in a cluster with the Dynamic Voltage Frequency Scaling (DVFS) technique. Takouna et. al. [24] presented a power-aware multi-core scheduling and their VM allocation algorithm selects a host which has the minimum increasing power consumption to assign a new VM. The VM allocation algorithm, however, is similar to the PABFDs [6] except that they are concerned about memory usage in a period of estimated runtime for estimating the host's energy. The work also presented a method to select optimal operating frequency for a (DVFS-enabled) host and configure the number of virtual cores for VMs. Our proposed EPOBF algorithms, which are VM allocation algorithms, differ from the these previous works. Our algorithms use the VM's starting time and finished time to minimize the total working time on physical servers, and consequently minimize the total energy consumption in all physical servers. In this paper, we do not use the DVFS-based technique to reduce energy consumption on a cloud data center. We propose software-based VM allocation algorithms which are independent of vendor-locked hardware. Moreover, our proposed EPOBFs finding method is different from these previous works and our EPOBFs finding method chooses which host has the highest ratio between total maximum of MIPS (in all cores) and the maximum value of power consumption.

Mämmelä et. al. [17] presented energy-aware First-In, First-Out (E-FIFO) and energy-aware Backfilling First-Fit and Best-Fit (E-BFF, E-BBF) scheduling algorithms for non-virtualized high performance computing system. The E-FIFO puts new job at the end of job-queue (and dequeue last), finds out an available host for the first job and turns off idle hosts. The E-BFF and E-BBF are similar to E-FIFO, but the E-BFF and E-BBF will attempt to assign jobs to all idle hosts. Unlike our proposed EPOBF, the Mämmelä's work does not consider power-aware VM allocation.

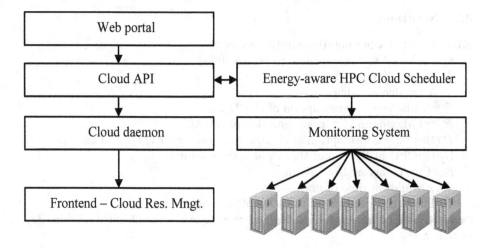

Fig. 1. Energy-Efficient HPC Cloud Architecture

3 Problem Description

3.1 Energy-Aware HPC Cloud Architecture

We descrive the Energy-Efficient HPC Cloud architecture in this section. The architecture includes Energy-aware HPC Cloud scheduler, monitoring system, cloud daemon, and web portal. This architecture is used to allocate resources to user HPC virtual machines in fulfillment of quality of service (e.g., performance and resource availability on requested time), while minimizing energy consumption.

- Energy-aware HPC Cloud scheduler uses energy-aware scheduling algorithms to provision VMs to user requests. The scheduler decides which physical machines to assign new VMs. In the following sections, we focus on problem of VM allocation and energy-aware algorithm named EPOBF to solve the problem.
- The monitoring system supervises the system resource usage, and reports the system resource usages to the scheduler.
- CloudAPI is a module that combines these components together in whole HPC Cloud system.
- Cloud daemon receives messages from CloudAPI to create templates that are compliant to a cloud resource management software (e.g. OpenNebula) to create VMs in the HPC Cloud system. Moreover, the Cloud daemon will setup environment for group of VMs by context generator. The group of VMs is used for parallel programs (e.g. MPI programs, Hadoop MapReduce programs).
- Web portal: Users interact with the HPC Cloud system via the web portal. The portal provides users functions for registration, log-in, requesting a VM or group of VMs, connecting to VMs via WebSSH.

3.2   Notations

We use the following notations in this paper:

vm_i: The ith virtual machine to be scheduled.

M_j: The jth physical server.

S: A feasible schedule.

P_j^{idle}: Idle power consumption of the M_j.

P_j^{max}: Maximum power consumption of the M_j.

$P_j(t)$: Power consumption of a single physical server (M_j) at a time point t.

$U_j(t)$: CPU utilization of the M_j at a time point t.

ts_i: Fixed starting time of vm_i.

dur_i: Duration time of vm_i.

T: The maximum time of the scheduling problem.

$r_j(t)$: Set of indexes of all VMs that are assigned to the physical machine M_j at time t.

3.3   Power Consumption Model

In this paper, we use the following energy consumption model proposed in [10] for a physical server M_j. The power consumption of each physical server, denoted as $P_j(.)$, is formulated as:

$$P_j(t) = P_j^{idle} + (P_j^{max} - P_j^{idle}) \times U_j(t) \tag{1}$$

The CPU utilization, denoted as $U_j(t)$, of the physical server at time t is formulated as:

$$U_j(t) = \sum_{c=1}^{PE_j} \sum_{i \in r_j(t)} \frac{mips_{i,c}}{MIPS_{j,c}} \tag{2}$$

The energy consumption of the server at time t is formulated as:

$$E_j = \int_0^{T_j} P_j(U_j(t))dt \tag{3}$$

where:

$U_j(t)$: CPU utilization of the server M_j at time t and $0 \leq U_j(t) \leq 1$.

PE_j : Number of processing elements (i.e. cores) of the physical server M_j.

$mips_{i,c}$: Allocated MIPS of the cth processing element to the vm_i by M_j.

$MIPS_{j,c}$: Total MIPS of the cth processing element on the M_j.

T_j: Working time of a physical server M_j. The T_j is defined as union of time intervals of all VMs that are allocated to a physical machine j^{th} at time t.

$$T_j = \bigcup_{vm_i \in r_{j,t}} [ts_i; ts_i + dur_i] \tag{4}$$

The union of two time intervals $[a;b]$ and $[c;d]$ is defined as: $[a;b] \cup [c;d] = \{x \in \mathbb{R} \mid x \in [a;b] \text{ or } x \in [c,d]\}$

3.4   Virtual Machine Allocation Problem Formulation

Given a set of virtual machines $V = \{vm_i | i = 1, 2, ..., n\}$ to be scheduled on a set of physical servers $M = \{M_j | j = 1, 2, ..., m\}$. Each VM is represented as a d-dimensional vector of demand resources, i.e. $vm_i = (x_{i,1}, x_{i,2}, ..., x_{i,d})$. Similarly, each physical machine is denoted as a d-dimensional vector of capacity resources, i.e. $M_j = (y_{j,1}, y_{j,2}, ..., y_{j,d})$. We consider types of resources such as processing element (PE), computing power (Million instruction per seconds - MIPS), physical memory (RAM), network bandwidth (BW), and storage. In addition, the virtual machine has life-cycle with starting time and finishing time, i.e., each vm_i is started at a fixed starting time (ts_i) and is neither preemptive nor migrated during its period time $(ts_i + dur_i)$.

A feasible schedule S indicates a successful mapping of all VMs to physical servers , i.e. $\forall i \in \{1, 2..., n\}, \exists j \in \{1, 2, ..., m\} : allocated(vm_i, M_j)$ where $allocated(vm_i, M_j)$ holds when vm_i is allocated on the physical server M_j.

We assume that every host M_j can run any virtual machine and the power consumption model $P_j(t)$ of the host M_j has a linear relationship with CPU utilization as described in Eq. (1).

The virtual machine scheduling problem is NP-hard, even if all physical servers are identical and all virtual machines are identical too, the scheduling is still NP-hard with $d \geq 1$ [19].

The goal is to find out a feasible schedule S that minimizes the total energy consumption of the cloud system, denoted as $\sum_{j=1}^{m} E_j$ in the equation (5) as following with $i \in \{1, 2, ..., n\}$, $j \in \{1, 2, ..., m\}$, $t \in [0; T]$. (In this paper we have not yet concerned on the energy consumption for other systems, such as electrical converters, cooling systems, and network systems).

$$\textbf{Minimize} \sum_{j=1}^{m} E_j \qquad (5)$$

where E_j with $j = 1, 2, ..., m$ is total energy consumption of a physical machine M_j as shown in Eq. 3.

The scheduling problem has the following (hard) constraints:

- Constraint 1: Each VM is run by a physical server (host) at any time.
- Constraint 2: VMs do not request any resource larger than total capacity resource of their hosts.
- Constraint 3: Let $r_j(t)$ be the set of VMs that are allocated onto a host M_j. The sum of total demand resources of these allocated VMs is less than or equal to total capacity of the resources of the M_j.

$$\forall c = 1, ..., d : \sum_{vm_i \in r_j(t)} x_{i,c} \leq y_{j,c} \qquad (6)$$

where:

- $x_{i,c}$ is resource of type c (e.g. CPU core, computing power, memory) requested by the vm_i (i=1,2,...,n).

- $y_{j,c}$ is capacity resource of type c (e.g. CPU core, computing power, memory) of the physical machine M_j ($j = 1, 2, ..., m$).

Based on the above observation, we propose our energy-aware algorithms as presented in the next section.

4   EPOBF Algorithm for Virtual Machine Allocation

Inspired by the Green500 lists idea [21] that uses a metric of performance-per-watt (FLOPS/watt) to rank energy efficiency of HPC systems, we raise questions: how can we use a similar metric (e.g. TotalMIPS/Watt) as a criterion for selecting a host to assign a new VM, and is total energy consumption of the whole system minimum?

We assume that if a host has more number of cores, then it will have more number of MIPS/Watt. The number of MIPS/Watt of a host is a ratio of total maximum of MIPS, which is sum of total maximum of MIPS of all hosts cores, to its maximum power consumption (P_{max}). The objective of our proposed work is energy efficiency. In this paper, we propose our energy-aware scheduling algorithm that denoted as the EPOBF-ST and EPOBF-FT. The EPOBF-ST and EPOBF-FT are best-fit decreasing heuristics to allocate a new VM to a physical server in such a way that has the maximum ratio between total computing power (in MIPS) and power consumption (in Watts).

EPOBF-ST and EPOBF-FT: The EPOBF-ST (EPOBF-FT) will sort list of VMs and order them by earliest start time first (earliest finishing time first), then the core EPOBF will assign a VM to a host that has the highest G value. For all $h \in M$, the G_h (denoted in pseudo-code is $G[h]$) can be calculated as a ratio of total maximum of MIPS of the host h (sum of total MIPS of all of cores) to maximum power consumption at 100% CPU utilization (P_{max}) of the host h. We called G the metric of performance-per-watt. In summary, the EPOBF

Algorithm 1. EPOBF-ST and EPOBF-FT: Heuristics for energy-aware VM allocation in HPC Clouds

---

```
 1: function EPOBF-ST
 2:     Input: vmList - a list of virtual machines to be scheduled
 3:     Input: hostList - a list of physical servers
 4:     Output: mapping (a feasible schedule) or null
 5:     vmList = sortVmListByStartTime( vmList )                        ▷ 1
 6:     return EPOBF( vmList, hostList )
 7: end function
 8: function EPOBF-FT
 9:     Input: vmList - a list of virtual machines to be scheduled
10:     Input: hostList - a list of physical servers
11:     Output: mapping (a feasible schedule) or null
12:     vmList = sortVmListByFinishTime( vmList )                       ▷ 2
13:     return EPOBF( vmList, hostList )
14: end function
```

Algorithm 2. EPOBF: Core EPOBF algorithm for energy-aware VM allocation in HPC Clouds

---

1: **function** EPOBF
2:     *Input:* vmList - a sorted list of virtual machines to be scheduled
3:     *Input:* hostList - a list of physical servers
4:     *Output:* mapping (a feasible schedule) or null
5:     **for** $vm \in vmList$ **do**
6:         host = FindHostForVmByGreenMetric(vm, hostList);
7:         CloudSim.allocationMap.put(vm.getId(),host.getId());
8:     **end for**
9:     Return mapping
10: **end function**
11: **function** FINDHOSTFORVMBYGREENMETRIC
12:     *Input:* vm - a virtual machine
13:     *Input:* hostList - a list of physical servers
14:     *Output:* bestHost - a best host for allocation of the *vm*
15:     bestHost = null
16:     maxG = 0
17:     CandidateHosts = findCandidateHosts(vm, hostList);
18:     **for** $h \in CandidateHosts$ **do**
19:         **if** h.checkAvailableResource(vm) **then**
20:             ▷ Available resources of the host h has enough the vm's requested resources
21:             **if** h.isOverUtilizedAfterAllocationVm(vm) **then**
22:                 ▷ host h is over utilized after allocation the *vm*
23:                 continue
24:             **end if**
25:             h.vmCreate(vm)   ▷ begin test
26:             G[h] = h.TotalMIPS / h.GetPower(1);
27:             h.vmDestroy(vm)   ▷ end test
28:             **if** (maxG < G[h]) **then**
29:                 maxG = G[h]
30:                 bestHost = h
31:             **end if**   ▷ Next iterate over set of candidate host
32:         **end if**
33:         **if** (bestHost != null) **then**
34:             allocate the *vm* to the *host*
35:             add the pair of *vm* (key) and *host* to the *mapping*
36:         **end if**
37:     **end for**   ▷ end for host list
38:     return *mapping*
39: **end function**

---

assigns each ready virtual machine v to the host h that has maximum $G[h]$ value. Algorithm 1 shows the pseudo-code for the EPOBF-ST and EPOBF-FT and Algorithm 2 shows the pseudo-code for the main core of the EPOBF.

In the Algorithm 1, the EPOBF-ST sorts all VMs by earliest starting time first (line 5) and the EPOBF-FT sorts all VMs by earliest finishing time first (line 12).

Table 1. Server characteristics (Type A: HP ProLiant ML110 G5, Type B: IBM x3250, Type C: Dell PowerEdge R620)

Server	Type A	Type B	Type C
CPU	1x Xeon 3075 with 2.66GHz	1x Xeon X3470 with 2.93GHz	2x Xeon E5-2660 with 2.20GHz
Number of cores (1)	2	4	16
Maximum of MIPS/core (2)	2660	2933	2660
Pidle (Unit: Watt)	93.7	41.6	56.1
Pmax (Unit: Watt) (3)	135.0	113.0	263.0
Total MIPS of all cores (4) = (1) * (2)	5,320.0	11,732.0	42,560.0
TotalMIPS/Pmax (5) = (4) / (3)	39.4	103.8	161.8
RAM (GB)	4	8	24
Network bandwidth (Kbits/s)	10,000,000	10,000,000	10,000,000

In the Algorithm 2, the EPOBF tries to allocate all VMs to physical machines. In the lines 5-8, it iterates on the list of VMs and then places the first VM to a physical machine h whose available resources can satisfy required resources of the VM and the $G[h]$ value of the physical machine h is maximum. It does so by invoking the FindHostForVmByGreenMetric function. At line 17, the findCandidateHosts(vm, H) function returns a set of candidate hosts which have sufficient available resources to satisfy all required resource constraints on the vm. Lines 18-37, the EPOBF chooses the best host h from the set of candidate hosts (denoted as $candidateHosts$), in which the host h has the maximum ratio of total host MIPS and maximum power host (denoted as G_h). The TotalMIPS is the maximum total MIPS of the host h, and the GetPower(1) function returns the maximum power consumption of the host h.

5   Performance Evaluation

This section presents the results obtained from our comparative simulated experiments. The simulated experiments aim to demonstrate and evaluate performance of the proposed energy-aware scheduling heuristics which are best-fit decreasing heuristics (denoted as EPOBF-ST and EPOBF-FT, in comparison with (1) PABFD, which is a popular power-aware VM allocation heuristic, (2) VBP Greedy L1 and L2 that are two norm-based vector bin packing heuristics. Our proposed EPOBF-ST and EPOBF-FT heuristics are presented in Section 4. The following section will present the PABFD and two norm-based vector bin packing (VBP Greedy L1 and VBP Greedy L2) heuristics.

PABFD: The PABFD (Power-Aware Best-Fit Decreasing), whose objective is minimizing total energy consumption of physical machines, uses as a baseline algorithm on problem of VM allocation for comparison in this paper. The PABFD sorts all VMs by their current requested CPU utilization from highest to lowest and then uses a Best-Fit Decreasing heuristic to select a physical machine for the first new virtual machine that has minimizing increasing power on each placement of the VM to assign a new VM. The PABFD is presented in [6] and also denoted as MBFD in [5].

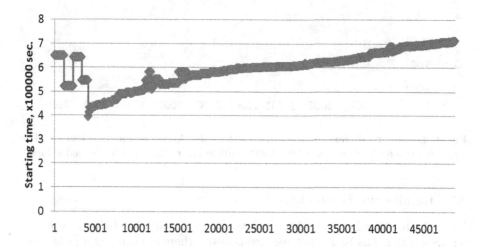

Fig. 2. The HPC simulated workload with starting times of 48,880 cloudlets

Table 2. Energy consumption (KWh)

VM Allocation Heuristic	Num. of Hosts	Num. of VMs	Energy (KWh)	Energy Saving (%) +: better; -:worst	Shutdown Hosts
PABFD	10000	48880	844.28	0%	21484
VBP Greedy L1	10000	48880	1072.86	-27%	23243
VBP Greedy L2	10000	48880	1072.86	-27%	23243
VBP Greedy L30	10000	48880	1072.86	-27%	23243
EPOBF-ST	10000	48880	664.16	21%	19025
EPOBF-FT	10000	48880	551.45	35%	18904

VBP greedy L1, *VBP greedy L2*, and *VBP greedy L30*: We implemented the vector bin packing (VBP) heuristic that is presented as Norm-based Greedy L1, Norm-based Greedy L2, and Norm-based Greedy L30 in [19]. All weights of the Norm-based Greedy heuristics are proposed in and are implemented as FFDAvgSum [19], which takes weights of resource dimensions that are exponential in average of the sum of demand resources.

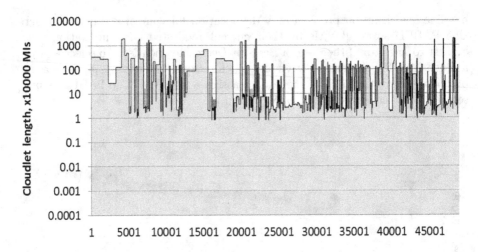

Fig. 3. Length in Millions Instructions (MI) of the 48,880 simulated cloudlet. The y-axis is the length of cloudlets (in x10000 Million Instructions) in log-scaled values.

5.1 Results and Discussions

We evaluate these heuristics by simulations using a simulated cloud data center system that has 10,000 heterogeneous hosts (there are three groups of hosts with different cores and physical memory), and some simulated HPC workload with thousands of CloudSim's cloudlets [9] (we assume that each HPC job's task is modeled as a cloudlet). The information of virtual machines (and also cloudlets) in these simulated workloads is extracted from a real log-trace (SDSC-BLUE-2000-4.1-cln.swf [3]) in Feitelson's Parallel Workloads Archive (PWA) [2] to model HPC jobs (original job's executing time adds 300 seconds that is over-head of using each VM). When converting from the log-trace, each cloudlet's length is a product of the system's processing time and CPU rating (we set the CPU rating is 375). We assign job's submission time, start time, and execution time, and number of processors in job data from the SDSC-BLUE-2000-4.1-cln to cloudlet's submission time, starting time and cloudlet's length, and number of cloudlets. We use four types of VMs (e.g. Amazon EC2's VM instances: high-CPU, extra, small, micro): each VM has only one core and maximum performance of VM is 2500, 2000, 1000, 500 MIPS, 870, 3840, 1740, 613 MB of RAM and network of 10000 Kbits/s. We choose the latest version (version 3.0) of the CloudSim [9] to model and simulate our HPC cloud and the VM allocation heuristics. The CloudSim's framework is used to write and evaluate new customized allocation algorithms for VMs and cloudlets. We based on the previous work [6] to improve existing PABFD allocation heuristic for this paper.

We show here the results of simulations using a simulated HPC workload with 48,880 cloudlets, in which parameters of all cloudlets are taken from 800 jobs in

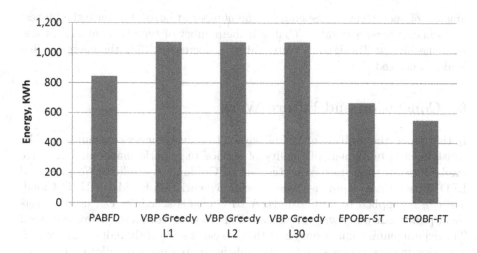

Fig. 4. Energy consumption (Unit: kWh)

the SDSC-BLUE-2000-4.1-cln workload as described above. Fig. 2 shows starting time of 48,880 cloudlets and Fig. 3 shows length in millions of instructions of the 48,880 cloudlets.

Our simulated Cloud datacenter has total 10,000 heterogeneous physical machines (PMs). These PMs include three groups of machines: one-third of HP ProLiant ML110 G5 machines, one-third of IBM x3250 machines, and one-third of Dell PowerEdge R620 machines. We assume that power consumption of a PM has a linear relationship to its CPU utilization (Equation 1). We use three power models of the three mainstream servers as summarized in Table 1 below. Table 1 shows server characteristics of three type of mainstream servers (Type A: HP ProLiant ML110 G5, Type B: IBM x3250, Type C: Dell PowerEdge R620).

The simulation results show here for VM allocation heuristics that have been presented in the previous section. Our scheduling objective is minimizing total energy consumption. Table 2 shows the energy consumption (kWh) of VM allocation heuristics. The Table 2 and Fig. 4 show scheduling results from simulations. Fig. 4 shows total energy consumption (kWh) of VM allocation heuristics: PABFD, VBP Greedy L1 and VBP Greedy L2, EPOBF-ST and EPOBF-FT In the Table 2, the percentages of energy savings of VBP Greedy L1 and VBP Greedy L2, EPOBF-ST and EPOBF-FT in comparison with the PABFD, if the energy savings of a heuristic is a positive number, then the heuristic is better than the PABFD. Otherwise, the heuristic is worse than the PABFD. The smaller number of shutdown hosts (column names as shutdown hosts) is better. Simulation results show that, our proposed EPOBF-FT is better than EPOBF-ST in energy saving (21% compared with 35%). A limitation of the EPOBF-ST and EPOBF-FT algorithms is that their amount of energy saving depends on the number of hosts that have the highest performance-per-watts ratio. That is if the number of type C servers (i.e. those with the highest TotalMIPS/Watts ratio) is decreased, then the energy saving is also decreased.

6   Conclusion and Future Work

In this paper, the problem of VM allocation to reduce energy consumption while satisfying the fulfillment of quality of service (e.g. performance and resource availability on time user requested) in HPC Cloud is studied. We presented EPOBF, which is a power-aware allocation heuristic of VMs in HPC Cloud, and can be applied to HPC clouds. A HPC clouds scheduler can use the metric of performance-per-watt to allocate VMs to hosts for more energy efficiency. The experimental simulations show that we can significantly reduce energy consumption in comparison with the state-of-the-art power-aware allocation heuristics (e.g. PABFD). The EPOBF-ST and EPOBF-FT heuristics could be a new VM allocation solution in a Cloud data center with heterogeneous and multi-core physical machines. Both versions of EPOBF-ST and EPOBF-FT heuristics are better than the PABFD and VBP greedy L1/L2/L30 allocation heuristics. The percentage of energy saving depends on how much energy consumption on server types and number of hosts that have the highest performance-per-watts ratio.

A limitation of our work is evaluating the performance of the EPOBF-ST and EPOBF-FT heuristics on various system models and workloads to provide a pros and cons of the EPOBF-ST and EPOBF-FT heuristics. Furthermore, we will consider limitations on computing resources and the impact of other components such as physical memory, network bandwidth in performance and energy consumption. In future, we plan to integrate the EPOBF-ST and EPOBF-FT heuristics into a cloud resource management software (e.g. Open-Stack Nova Scheduler). The cloud systems can provide resources to many types of VM-based leases [22] including best-effort, advanced reservation, and immediate leases at the same time. We are also studying Mixed Integer Linear Programming models and meta-heuristics (e.g. Genetic Algorithms) of the VM allocation problem.

References

1. AWS - High Performance Computing - HPC Cloud Computing. http://aws.amazon.com/hpc/ (retrieved on August 31, 2014)
2. Parallel Workloads Archive. http://www.cs.huji.ac.il/labs/parallel/workload/ (retrieved on January 31, 2014)
3. SDSC-BLUE-2000-4.1-cln.swf.gz log-trace. http://www.cs.huji.ac.il/labs/parallel/workload/l_sdsc_blue/SDSC-BLUE-2000-4.1-cln.swf.gz (retrieved on Januray 31, 2014)

4. Albers, S.: Energy-efficient algorithms. Commun. ACM **53**(5), 86–96 (2010)
5. Beloglazov, A., Abawajy, J., Buyya, R.: Energy-aware resource allocation heuristics for efficient management of data centers for cloud computing. Future Generation Comp. Syst. **28**(5), 755–768 (2012)
6. Beloglazov, A., Buyya, R.: Optimal online deterministic algorithms and adaptive heuristics for energy and performance efficient dynamic consolidation of virtual machines in cloud data centers. Concurrency and Computation: Practice and Experience **24**(13), 1397–1420 (2012)
7. Beloglazov, A., Buyya, R., Lee, Y.C., Zomaya, A.: A Taxonomy and Survey of Energy-Efficient Data Centers and Cloud Computing Systems. Advances in Computers **82**, 1–51 (2011)
8. Buyya, R., Yeo, C., Venugopal, S., Broberg, J., Brandic, I.: Cloud computing and emerging it platforms: Vision, hype, and reality for delivering computing as the 5th utility. Future Generation Comp. Syst. **25**(6), 599–616 (2009)
9. Calheiros, R.N., Ranjan, R., Beloglazov, A., De Rose, C.A.F., Buyya, R.: Cloudsim: A toolkit for modeling and simulation of cloud computing environments and evaluation of resource provisioning algorithms. Softw., Pract. Exper. **41**(1), 23–50 (2011)
10. Fan, X., Weber, W.D., Barroso, L.: Power provisioning for a warehouse-sized computer. In: ISCA, pp. 13–23 (2007)
11. Garg, S.K., Yeo, C.S., Anandasivam, A., Buyya, R.: Energy-Efficient Scheduling of HPC Applications in Cloud Computing Environments. CoRR abs/0909.1146 (2009)
12. Goiri, I., Julia, F., Nou, R., Berral, J.L., Guitart, J., Torres, J.: Energy-Aware Scheduling in Virtualized Datacenters. In: 2010 IEEE International Conference on Cluster Computing, pp. 58–67. IEEE (September 2010). http://doi.ieeecomputersociety.org/10.1109/CLUSTER.2010.15. http://ieeexplore.ieee.org/lpdocs/epic03/wrapper.htm?arnumber=5600320
13. Jing, S.Y., Ali, S., She, K., Zhong, Y.: State-of-the-art research study for green cloud computing. The Journal of Supercomputing 65(1), 445–468 (2013). http://www.springerlink.com/index/10.1007/s11227-011-0722-1. http://link.springer.com/10.1007/s11227-011-0722-1
14. von Laszewski, G., Wang, L., Younge, A.J., He, X.: Power-aware scheduling of virtual machines in dvfs-enabled clusters. In: CLUSTER, pp. 1–10 (2009)
15. Le, K., Bianchini, R., Zhang, J., Jaluria, Y., Meng, J., Nguyen, T.D.: Reducing electricity cost through virtual machine placement in high performance computing clouds. In: SC, p. 22 (2011)
16. Liu, Y., Zhu, H.: A survey of the research on power management techniques for high-performance systems. Software: Practice and Experience 40(11), 943–964 (2010). http://onlinelibrary.wiley.com/doi/10.1002/spe.952/abstract. http://onlinelibrary.wiley.com/doi/10.1002/spe.952/pdf. http://cms.brookes.ac.uk/staff/HongZhu/Publications/Power_Mgt-final.pdf
17. Mämmelä, O., Majanen, M., Basmadjian, R., de Meer, H., Giesler, A., Homberg, W.: Energy-aware Job Scheduler for High-performance Computing (2012)
18. Mauch, V., Kunze, M., Hillenbrand, M.: High performance cloud computing. Future Generation Comp. Syst. **29**(6), 1408–1416 (2013)
19. Panigrahy, R., Talwar, K., Uyeda, L., Wieder, U.: Heuristics for Vector Bin Packing. Tech. rep., Microsoft Research (2011)
20. Pham, T.V., Jamjoom, H., Jordan, K.E., Shae, Z.Y.: A service composition framework for market-oriented high performance computing cloud. In: HPDC, pp. 284–287 (2010)

21. Sharma, S.: Making a case for a green500 list, pp. 12–8 (2006)
22. Sotomayor, B.: Provisioning Computational Resources Using Virtual Machines and Leases. Ph.D. thesis, University of Chicago (2010)
23. Sotomayor, B., Keahey, K., Foster, I.T.: Combining batch execution and leasing using virtual machines. In: HPDC, pp. 87–96 (2008)
24. Takouna, I., Dawoud, W., Meinel, C.: Energy Efficient Scheduling of HPC-jobs on Virtualize Clusters using Host and VM Dynamic Configuration. Operating Systems Review **46**(2), 19–27 (2012)
25. Viswanathan, H., Lee, E.K., Rodero, I., Pompili, D., Parashar, M., Gamell, M.: Energy-Aware Application-Centric VM Allocation for HPC Workloads. In: IPDPS Workshops, pp. 890–897 (2011)

Human Object Classification Using Dual Tree Complex Wavelet Transform and Zernike Moment

Manish Khare[1], Nguyen Thanh Binh[2(✉)], and Rajneesh Kumar Srivastava[1]

[1] Department of Electronics and Communication, University of Allahabad, Allahabad, India
{mkharejk,rkumarsau}@gmail.com
[2] Faculty of Computer Science and Engineering, Ho Chi Minh City
University of Technology, Ho Chi Minh, Vietnam
ntbinh@cse.hcmut.edu.vn

Abstract. Presence of variety of objects degrade the performance of video surveillance system as a certain type of objects can be misclassified as some other types of object. Recent researches in video surveillance are focused on accurate classification of human objects. Classification of human objects is a crucial problem, as accurate human object classification is a desirable task for better performance of video surveillance system. In this paper we have proposed a method for human object classification, which classify the objects present in a scene into two classes: human and non-human. The proposed method uses combination of Dual tree complex wavelet transform and Zernike moment as feature of object. We have used support vector machine (SVM) as a classifier for classification of objects. The proposed method has been tested on standard dataset like INRIA person dataset. Quantitative experimental results shows that the proposed method is better than other state-of-the-art methods and gives better performance for human object classification.

Keywords: Object classification · Zernike moment · Dual tree complex wavelet transform · Support vector machine

1 Introduction

Human object classification in real scenes is a challenging problem with many useful applications like object tracking, object identification, etc. [1-2]. Goal of any object classification algorithm is to develop a method having capability to interpret the objects into different groups. Object classification algorithm must work under real time constraints and robust in the situations like, color variation in human cloths, variation in natural conditions, different size of human objects, etc. Feature selection and machine learning techniques are essential components of any classification algorithm [3]. Object classification algorithm is commonly divided into three components: (i) selection of feature, (ii) extraction of feature, and (iii) classification. Correctness of any classification scheme lies on the selected feature, therefore selection of effective feature is a crucial step for successful classification. Machine learning based methods have been used for developing object classification algorithm. The learning in a classification system is

© Springer-Verlag Berlin Heidelberg 2014
A. Hameurlain et al. (Eds.): TLDKS XVI, LNCS 8960, pp. 87–101, 2014.
DOI: 10.1007/978-3-662-45947-8_7

either based on pixel or on feature [4]. The feature based learning is beneficial than pixel based learning, because it is difficult to train finite quantity of data using pixel in comparison to encoding of ad-hoc knowledge using feature.

To address the human object classification, various researchers proposed their solutions. Some approaches are based on pixel based method whereas some approaches are based on feature based method. Sialet et al. [5] used Haar like features along with the decision tree in their pedestrian detection system. Viola and Jones [6] used modified version of Haar basis function for object detection. Dalal and Triggs [7] proposed Histogram of Oriented Gradient (HoG) as a feature descriptor for object detection. Cao et al. [8] proposed a method by extending the Histogram of oriented Gradient, to boost the HoG features. Lu and Zheng [9] proposed a visual feature for object classification based on binary pattern. These visual features are rotation invariant and exploits the property of pixel pattern. Lowe [10] used Scale Invariant Feature Transform (SIFT) as a feature descriptor for object recognition. All the methods discussed above depend on only one feature evaluation set therefore they have some local advantages depending on the features used.

Yu and Slotine [11] proposed a wavelet based method for classification. Method proposed by Yu and Slotine [11] uses real valued wavelet transform, but real valued wavelet transform is not suitable for surveillance application, because in case of video surveillance, object of interest may be present in translated or rotated form among different frames and coefficients of real valued wavelet transform corresponding to object region changes abruptly across different frame [12,13]. Use of complex wavelet transform can avoid this shortcoming as it is shift invariant in nature.

Motivated by work of Yu and Slotine and properties of complex wavelet transform Khare et al. [2] proposed human object classification method by using Dual tree complex wavelet transform. Combining two or more features in one is the recent trend and give more accurate results in comparison with use of single feature based object classification. Here we are extending our earlier work [2], by proposing a new method for human object classification based on combination of Dual tree complex wavelet transform and Zernike moment as a feature set. We have used support vector machine (SVM) classifier for classifying human and non-human object classes. The Dual tree complex wavelet transform having advantages of shift invariance and better edge representation as compared to real valued wavelet transform. Zernike moment also have many important properties such as translation invariance, rotation invariance etc.

We have experimented the proposed method at multiple levels of Dual tree complex wavelet transform. We have also compared the proposed method with the method using coefficients of real-valued discrete wavelet transform as a feature set. We have compared the proposed method with other state-of-the-art methods proposed by Khare et al. [2], Dalal and Triggs [7], Lu and Zheng [9], Renno et al. [14], and Chen et al. [15] in terms of confusion matrix. We have taken three different performance metrics: average classification accuracy, true positive rate (recall), and predicted positive rate (precision).

Rest of paper is organized as follows: Section 2 describes basics of features (Dual tree complex wavelet transform and Zernike moment). Section 3 describes support vector machine classifier and section 4 describes the proposed method. Experimental results, analysis and comparison of the proposed method with other state-of-the-art methods are given in section 5. Finally conclusions of the work are given in section 6.

2   Feature Selection

Castleman [16] defined feature as, "A feature is a function of one or more measurement computed so that it quantifies some significant characteristics of object". In any object classification algorithm, selection of appropriate feature is very important. If corrected feature is selected for classification algorithm then performance of the classifier will increase. Use of single type of feature may not be sufficient for solving human object classification problem, because single type of feature is not rich enough for representation of different shades of human objects. Combining multiple type of features can enhance the robustness and classification accuracy of the classifier for human object classification. When we use combination of two or more features, some of the features are more informative than others for a particular case, therefore chances of correct classification will be high. In the proposed work for human object classification, we have taken combination of two different features – Dual tree complex wavelet transform (DTCWT) and Zernike moment (ZM). A brief description of these two features and why they are useful for human object classification are described in subsection 2.1 and 2.2 respectively.

2.1   Dual Tree Complex Wavelet Transform

Real valued wavelet transform suffers from two major shortcomings - lack of shift sensitivity and lack of strong edge detection. Kingsbury et al. [17,18] proposed a solution to overcome these problems in form of use of Dual Tree Complex Wavelet Transform (DT-CWT). DT-CWT is shift invariant in nature and gives strong edge information for filtering multidimensional signals. Dual tree complex wavelet transform have following properties as described in [17, 18].

- Approximate shift invariance.
- Good directional selectivity in *2-D* with Gabor like filters (Also true for *m-dimensional*).
- Perfect reconstruction using short linear phase filters.
- Limited Redundancy, independent of the number of scales (*2:1* for *1-Dimensional* and $2^n:1$ for *n-Dimensional*).
- Efficient order N computation – only double of the simple DWT for *1-D* (2^n times for *n-dimensional*).

Human object classification is a problem where the objects may present in translated as well as rotated form among different scenes. Therefore first three properties of dual tree complex wavelet transform will be useful for classification of human object.

For implementation of DT-CWT, Kingsbury [18] analyzed that the approximate shift invariance with a real DWT can be achieved by doubling the sampling rate at each level of the wavelet decomposition tree. For this work, the samples must be evenly spaced. One approach to double the sampling rate in Tree a of Fig. 1(a) is by eliminating the down sampling by 2 after the level 1 filters, H_{0a} and H_{1a}. This is equivalent to two parallel fully decimated tree a and b in Fig. 1(a), provided that the decays of H_{0b} and H_{1b} are one sample offset from H_{0a} and H_{1a} which ensure that the level *1* down samplers in tree *b* pick the opposite sample to those in tree *a*. then it is found that to get uniform intervals between samples from the two trees below level *1*, the filter in one tree must provide delays that one half of a sample different from those in the opposite tree.

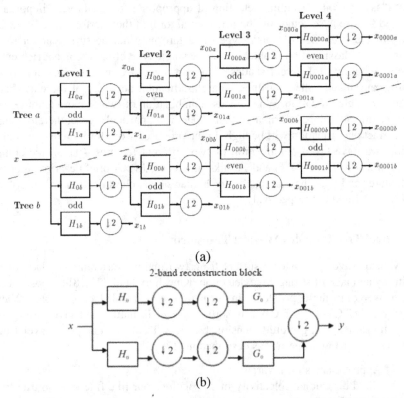

(a)

(b)

Fig. 1. Dual tree of the real filters for the complex wavelet transform (DTCWT), giving real and imaginary parts of the complex coefficients

For inverse of Dual tree complex wavelet transform each tree of Fig. 1(a) is inverted separately using biorthogonal filter *G*, specially designed for perfect reconstruction with analysis filter *H* in the *2-band* reconstruction block in Fig. 1(b). The results of each of the 2-inverse function are averaged to give reconstructed signal. This system is a wavelet frame with redundancy two; and if the filters are designed such that the analysis and re-construction filters have very similar frequency responses, then it is an almost tight frame, which means that energy is approximately preserved when signals are transformed into the DT-CWT domain.

2.2   Zernike Moment

Zernike moment was firstly introduced by Teague [19] to overcome the shortcomings of information redundancy present in Geometric moments [19]. Zernike moment can represent the properties of an image with no redundancy or overlap of information between the moments [20]. Due to these characteristics, Zernike moment have been utilized as feature set in different applications such as object classification, shape analysis, content based image retrieval etc.

Zernike moment (ZM) owns following properties [21,22]:

(i)   Zernike moments are rotation, translation and scale invariant [23].
(ii)   Zernike moments are robust to noise and minor variation in shape.
(iii) Since the basis of Zernike moment is orthogonal, therefore they have minimum information redundancy.
(iv)   Zernike moment can characterize the global shape of pattern. Lower order moments represent the global shape pattern and higher order moment represents the detail.
(v)   An image can be better described by a small set of its Zernike moments than any other types of moments.

Zernike moment is a set of complex polynomial which form a complete orthogonal set over the interior of the unit circle of $x^2 + y^2 \leq 1$ [24,25]. These polynomials are of the form,

$$V_{mn}(x, y) = V_{mn}(r, \theta) = R_{mn}(r).\exp(jn\theta) \tag{1}$$

where m is positive integer and n is integer subject to constraints $m-|n|$ even and $|n| \leq m$, r is the length of vector from the origin to pixel (x,y) and θ is the angle between vector r and x-axis in counter clock wise direction, $R_{mn}(r)$ is the Zernike radial polynomials in (r,θ) polar coordinates and defined as

$$R_{mn}(r) = \sum_{s=0}^{\left(\frac{m-|n|}{2}\right)} \frac{(-1)^s (m-s)! r^{m-2s}}{s!\left(\frac{m+|n|}{2} - s\right)!\left(\frac{m-|n|}{2} - s\right)!} \tag{2}$$

here $R_{m,-n}(r) = R_{mn}(r)$

The above mentioned polynomial in equation (2) is orthogonal and satisfies the othogonality principle

Zernike moments are the projection of image function $I(x,y)$ onto these orthogonal basis function. The Othogonality condition simplifies the representation of the original image because generated moments are independent [26].

The Zernike moment of order m with repetition n for a continuous image function $I(x,y)$ that vanishes outside the unit circle is

$$Z_{mn} = \frac{m+1}{\pi} \iint\limits_{x^2+y^2 \leq 1} I(x, y)[V_{mn}(r,\theta)]dxdy \tag{3}$$

In case of digital image, the integral are replaced by summation, given as follows

$$Z_{mn} = \frac{m+1}{\pi} \sum_{x} \sum_{y} I(x, y) V_{mn}(r, \theta), \quad x^2 + y^2 \le 1 \qquad (4)$$

Dual tree complex wavelet transform and Zernike moments both have its own properties such as: shift-invariant [27], translation invariant [22], rotation invariant [23] etc. So if we use combination of these two features in one methodology then we can expect for more accurate object classification results in comparison to use of only Dual tree complex wavelet transform or Zernike moment.

3   Support Vector Machine Classifier

Support vector machine (SVM) is very popular classifier, which classifies objects into two categories: object and non-object data [28]. An n-dimensional object x has n-coordinates.

$$x = (x_1, x_2, x_3, \ldots, x_n) \text{, where, each } x_i \in R \text{ for } i = 1, 2, 3, \ldots, n.$$

Each object x_j belongs to a class $y_j \in \{-1, +1\}$. Consider a training set T of m patterns together with their classes, $T = \{(x_1, y_1), (x_2, y_2), \ldots, (x_n, y_n)\}$ and a dot product space S, in which the objects are embedded, $x_1, x_2, \ldots, x_m \in S$. Any hyperplane in the space S can be written as

$$\{x \in S \mid w.x + b = 0\}, w \in S, b \in R \qquad (5)$$

The dot product w.x is defined by:

$$w.x = \sum_{i=1}^{n} w_i x_i \qquad (6)$$

A training set of objects is linearly separable if there exists at least one linear classifier defined by the pair *(w, b)* which correctly classifies all objects. The linear classifier is represented by the hyperplane *H (w.x+b=0)* and defines a region for class *+1* and another region for class *-1* objects. After training, the classifier is ready to predict the class membership for new objects, different from those used in training. The class of object x_k is determined with the equation:

$$class(x_k) = \begin{cases} +1 & if \quad w.x_k + b > 0 \\ -1 & if \quad w.x_k + b < 0 \end{cases} \qquad (7)$$

4   The Proposed Method

The paper presents a new method for human object classification, a variation of method which was published by Khare et al. [2]. In present paper, we extended our earlier work [2], by introducing combination of Dual tree complex wavelet transform and Zernike moment as a feature evaluation set. The proposed method uses support vector machine as a classifier for classification of data into two categories: human object and non-human object. Steps of the proposed method are described below-

Step 1: First step of the proposed method is to collect sample images for training and testing the classifier. In the present work, we have taken INRIA person dataset [31] images for training and testing purpose. The INRIA person dataset [29] contains 2521 positive images and 1686 negative images. Positive images contain human object images and negative images contain any type of images in which human object is not present. We have taken 500 images of each category (positive images and negative images respectively) for training purpose and 100 images of each category (positive images and negative images respectively) for testing purpose.

Step 2: INRIA person dataset images are of different size, so normalization of these images are required in order to reduce complexity. In the present work, the collected images are scale normalized to 256 x 256 pixel dimensions. After normalization, the scaled normalized images of the RGB color space were converted to the gray level images.

Step 3: Third step of the proposed method is feature vector computation. For feature vector computation in the proposed method, image frames are decomposed into complex wavelet coefficients using Dual tree complex wavelet transform. After applying Dual tree complex wavelet transform, we get coefficients in form of two filters: low-pass filter image and high-pass filter image (as shown in Fig. 1(A)). Values of high-pass filtered image are used as feature values of different images, because high-pass filtered image provides detail coefficient of images, which is in form of complex value. We have skipped values of low-pass filtered image, because low-pass filter image provides approximation coefficients of image. We applied Zernike moment of 4th order in high-pass filter image wavelet coefficients. Values obtained after Zernike moment computation over dual tree complex wavelet coefficients have been used as feature set.

Step 4: Fourth step of the proposed method is to train the classifier using feature values, which we have obtained in step 3. We have used SVM classifier, in which we assign value '1' for human object and value '0' for non-human object. By using feature values of images and assignment values of data, SVM classifier trained for classification. Detailed information of SVM classifier is given in section 3.

Step 5: Final step of the proposed method is to classify the test data into one of the two categories: human and non-human object. For this process first we compute feature vector of image using step 3 of the proposed method, then this computed feature value is supplied into SVM classifier, where SVM classifier analyzes this feature value by previously trained data and gives result in form of two value '0' and '1', where '0' indicates non-human object data and '1' indicates human object data. Same process will be repeated for all test data.

5   Experimental Results

This section provides quantitative experimental results of the proposed method and other state-of-the-art methods proposed by Khare et al. [2], Dalal and Triggs [7], Lu and Zheng [9], Renno et al. [14], and Chen et al. [15]. We have also evaluated human object classification results by using discrete wavelet transform as a feature. Comparison of the results obtained by the proposed method and other state-of-the-art method has been done in terms of confusion matrix, average classification accuracy true positive rate (recall), and predicted positive rate (precision). The proposed method for human object classification has been tested on INRIA person dataset [29]. We have shown some human and non-human representative images of INRIA person dataset in Fig. 2. By observing the human object image of INRIA person dataset from Fig. 2(a), one can observe that both frontal as well as side view of human object were taken for experiments.

(a)

(b)

Fig. 2. Sample images with human and non-human objects of INRIA dataset

For experimentation, we have taken three types of features: (i). only Zernike moment, (ii). Only Dual tree complex wavelet transform coefficients (Khare et al. [2]), and (iii). Combination of Dual tree complex wavelet transform coefficients and Zernike moment coefficients. We have evaluated the proposed method for multiple levels of Dual tree complex wavelet transform coefficients (L − 1,2,….,7). Confusion matrix for the proposed method applied by using these three above mentioned feature set are given in Table 1, 2 and 3 respectively. Just to compare performance of the proposed method, we have also experimented with multilevel discrete wavelet transform coefficients as a feature set, for human object classification, and confusion matrix is given in Table 4. We have also evaluated the confusion matrix for other state-of-the-art methods proposed by Dalal and Triggs [7], Lu and Zheng [9], Renno et al. [14] and Chen et al. [15], and confusion matrix for them are given in Table 5.

Table 1. Confusion matrix for the proposed method using Zernike moment as a feature set

Zernike moment as a feature set		
True classes	Predicted classes	
	Human	Non-human
Human	81	19
Non-human	27	73

Table 2. Confusion matrix for the proposed method using Dual tree complex wavelet transform (DTCWT) coefficients at multilevel as a feature set [Khare et al. [2]]

DTCWT coefficients (Level – 1) as a feature set			DTCWT coefficients (Level – 2) as a feature set		
True classes	Predicted classes		True Classes	Predicted classes	
	Human	Non-human		Human	Non-human
Human	90	10	Human	92	8
Non-human	10	90	Non-human	10	90
DTCWT coefficients (Level – 3) as a feature set			DTCWT coefficients (Level – 4) as a feature set		
True classes	Predicted classes		True classes	Predicted classes	
	Human	Non-human		Human	Non-human
Human	95	5	Human	95	5
Non-human	6	94	Non-human	6	94
DTCWT coefficients (Level – 5) as a feature set			DTCWT coefficients (Level – 6) as a feature set		
True classes	Predicted classes		True classes	Predicted classes	
	Human	Non-human		Human	Non-human
Human	97	3	Human	99	1
Non-human	6	94	Non-human	3	97
DTCWT coefficients (Level – 7) as a feature set					
True classes	Predicted classes				
	Human		Non-human		
Human	99		1		
Non-human	1		99		

From Table 3, we found that the proposed method gives better human object classification result at higher levels of Dual tree complex wavelet transform coefficients after combining Zernike moment instead of only Dual tree complex wavelet transform coefficients as shown in Table 2 (Khare et al. [2] method results). From Table 4, we found that by using only discrete wavelet transform as feature, human objects have more confusion with non-human objects as compared with the results obtained by the proposed method. Confusion matrix result for single feature set as given in Table 1, 2 and 4 gives poor results in comparison with combination of two feature set results as given in Table 3. This shows that combining two features into one gives better results than use of single feature. Results shown in Tables 1-5, clearly demonstrate that the proposed method gives better human object classification results at higher levels of Dual tree complex wavelet transform in comparison to other state-of-the-art methods proposed by Dalal and Triggs [7], Lu and Zheng [9], Renno et al. [14] and Chen et al. [15].

We have also computed performance of the proposed method and all other state-of-the-art methods in terms of three performance metrics – average classification accuracy, True positive rate (also known as Recall), and Predicted positive rate (also known as Precision). For computation of these measures we have used four terms: TP (true positive), TN (true negative), FP (false positive), FN (false negative), which are defined as: TP are total number of images which are originally positive and detected as positive images, TN are total number of images which are originally negative and detected as negative images, FP are total number of images which are originally negative and detected as positive images, and FN are total number of images which are originally positive and detected as negative images.

Table 3. Confusion matrix for the proposed method using combination of Dual tree complex wavelet transform (DTCWT) coefficients at multilevel and Zernike moment coefficients as a feature set

Combination of DTCWT coefficients (level – 1) and Zernike moment coefficient as a feature set			Combination of DTCWT coefficients (level – 2) and Zernike moment coefficient as a feature set		
True classes	Predicted classes		True Classes	Predicted classes	
	Human	Non-human		Human	Non-human
Human	95	5	Human	95	5
Non-human	10	90	Non-human	7	93
Combination of DTCWT coefficients (level – 3) and Zernike moment coefficient as a feature set			**Combination of DTCWT coefficients (level – 4) and Zernike moment coefficient as a feature set**		
True classes	Predicted classes		True classes	Predicted classes	
	Human	Non-human		Human	Non-human
Human	96	4	Human	96	4
Non-human	7	93	Non-human	5	95
Combination of DTCWT coefficients (level – 5) and Zernike moment coefficient as a feature set			**Combination of DTCWT coefficients (level – 6) and Zernike moment coefficient as a feature set**		
True classes	Predicted classes		True classes	Predicted classes	
	Human	Non-human		Human	Non-human
Human	99	1	Human	99	1
Non-human	3	97	Non-human	2	98

Combination of DTCWT coefficients (level – 7) and Zernike moment coefficient as a feature set		
True classes	Predicted classes	
	Human	Non-human
Human	100	0
Non-human	2	98

Table 4. Confusion matrix for the proposed method using discrete wavelet transform (DWT) coefficients at multilevel as a feature set

DWT coefficients (Level – 1) as a feature set			DWT coefficients (Level – 2) as a feature set		
True classes	Predicted classes		True Classes	Predicted classes	
	Human	Non-human		Human	Non-human
Human	85	15	Human	85	15
Non-human	30	70	Non-human	30	70
DWT coefficients (Level – 3) as a feature set			**DWT coefficients (Level – 4) as a feature set**		
True classes	Predicted classes		True classes	Predicted classes	
	Human	Non-human		Human	Non-human
Human	89	11	Human	91	9
Non-human	27	73	Non-human	21	79
DWT coefficients (Level – 3) as a feature set			**DWT coefficients (Level – 4) as a feature set**		
True classes	Predicted classes		True classes	Predicted classes	
	Human	Non-human		Human	Non-human
Human	93	7	Human	95	5
Non-human	20	80	Non-human	11	89

DWT coefficients (Level – 7) as a feature set		
True classes	Predicted classes	
	Human	Non-human
Human	95	5
Non-human	11	89

Table 5. Confusion matrix for the state-of-the-art methods [7,9,14,15]

Method proposed by Dalal and Triggs [7]			Method proposed by Lu and Zheng [9]		
True classes	Predicted classes		True Classes	Predicted classes	
	Human	Non-human		Human	Non-human
Human	96	04	Human	90	10
Non-human	08	92	Non-human	30	70
Method proposed by Renno et al. [14]			**Method proposed by Chen et al. [15]**		
True classes	Predicted classes		True classes	Predicted classes	
	Human	Non-human		Human	Non-human
Human	89	11	Human	98	02
Non-human	31	69	Non-human	04	96

Average classification accuracy is defined as the proportion of the total number of prediction that were correct. Average classification accuracy can be calculated using following formula.

$$Average\,Classification\,Accuracy = \frac{TP + TN}{TP + TN + FP + FN} \tag{8}$$

True positive rate (TPR) (also referred as Recall) is defined as the proportion of positive cases that were correctly classified as positive. TPR can be calculated using following formula.

$$TPR(\text{Re}\,call) = \frac{TP}{TP + FN} \qquad (9)$$

Predicted positive rate (PPR) (also referred as Precision) is defined as the proportion of the predicted positive cases that were correct. PPR can be calculated using following formula.

$$PPR(\text{Pr}\,ecision) = \frac{TP}{FP + TP} \qquad (10)$$

Values of all performance measures have been shown in Table 6, for the proposed method and all other state-of-the-art methods.

Table 6. Performance measure values

Methods Name	TPR (Recall) (%)	PPR (Precision) (%)	Average Classification Accuracy (%)
The Proposed method with Zernike Moment as a feature	81.00	75.00	77.00
The Proposed method with DTCWT (Level-1) as a feature (Khare et al. [2])	90.00	90.00	90.00
The Proposed method with DTCWT (Level-2) as a feature (Khare et al. [2])	92.00	90.19	91.00
The Proposed method with DTCWT (Level-3) as a feature (Khare et al. [2])	95.00	94.06	94.50
The Proposed method with DTCWT (Level-4) as a feature (Khare et al. [2])	95.00	94.06	94.50
The Proposed method with DTCWT (Level-5) as a feature (Khare et al. [2])	97.00	94.17	95.50
The Proposed method with DTCWT (Level-6) as a feature (Khare et al. [2])	97.00	97.06	98.00
The Proposed method with DTCWT (Level-7) as a feature (Khare et al. [2])	99.00	99.00	99.00
The Proposed method with combination of DTCWT (Level-1) and Zernike moment as a feature	95.00	90.47	92.50
The Proposed method with combination of DTCWT (Level-2) and Zernike moment as a feature	95.00	93.12	94.00
The Proposed method with combination of DTCWT (Level-3) and Zernike moment as a feature	96.00	93.21	94.50

Table 6. (*continued*)

Methods Name	TPR (Recall) (%)	PPR (Precision) (%)	Average Classification Accuracy (%)
The Proposed method with combination of DTCWT (Level-4) and Zernike moment as a feature	96.00	95.05	95.50
The Proposed method with combination of DTCWT (Level-5) and Zernike moment as a feature	97.00	97.00	97.00
The Proposed method with combination of DTCWT (Level-6) and Zernike moment as a feature	99.00	98.02	98.50
The Proposed method with combination of DTCWT (Level-7) and Zernike moment as a feature	100.00	98.04	99.00
DWT (Level-1) as a feature	85.00	73.91	77.50
DWT (Level-2) as a feature	85.00	73.91	77.50
DWT (Level-3) as a feature	89.00	76.92	81.00
DWT (Level-4) as a feature	91.00	81.25	85.00
DWT (Level-5) as a feature	93.00	82.30	86.60
DWT (Level-6) as a feature	95.00	89.62	92.00
DWT (Level-7) as a feature	95.00	89.62	92.00
Method Proposed by Dalal and Triggs [7]	96.00	92.31	94.00
Method Proposed by Lu and Zheng [9]	90.00	75.00	80.00
Method Proposed by Renno et al. [14]	89.00	74.17	79.00
Method Proposed by Chen et al. [15]	98.00	96.08	97.00

From Table 6, one can observe that the proposed method with combination of Dual tree complex wavelet transform coefficients and Zernike moment as a feature set gives better performance at higher levels of Dual tree complex wavelet transform in comparison to other state-of-the-art methods [7,9,14,15], and multilevel discrete wavelet transform, as feature, for human object classification in terms of three different quantitative performance measures –Average Classification Accuracy, TPR, and PPR.

6   Conclusions

In this paper, our goal is to develop and demonstrate a new method for human object classification in real scenes using combination of two feature set namely – Dual tree complex wavelet transform coefficients and Zernike moment. Dual tree complex wavelet transform is advantageous over real valued wavelet transform in terms of better edge representation and shift invariant property. Zernike moment is translation and rotation invariant. The proposed method for human object classification is an extension of work proposed by Khare et al. [2]. The proposed method first trains SVM classifier by using selected feature set, then classify new object data (testing data) into one of the two categories viz. human objet and non-human object. The proposed method is compared with other state-of-the-art methods proposed by Khare et al. [2], Dalal and Triggs [7], Lu and Zheng [9], Renno et al. [14], and Chen et al. [15] as well as classification results obtained by using Zernike moment as a feature set, classification results obtained by using discrete wavelet transform as a feature set. Comparison has been done in terms of average classification accuracy, TPR (Recall), and PPR (Precision). Quantitative experimental analysis shows that the proposed method gives better classification results than other discussed methods. Main advantage of the proposed method is that: the proposed method can detect human objects in complex background, as well as the proposed method can detect human objects of different size.

Acknowledgement. This work supported in part by Council of Scientific and Industrial Research (CSIR), Human Resource Development Group, India, Under Grant No. 09/001/(0377)/2013/EMR-I.

References

1. Hu, W., Tan, T.: A survey on visual Surveillance of object motion and behaviors. IEEE Transaction on System, Man and Cybernetics **34**(3), 334–352 (2006)
2. Khare, M., Binh, N.T., Srivastava, R.K.: Dual tree complex wavelet transform based human object classification using support vector machine. Journal of Science and Technology **51**(4B), 134–142 (2013)
3. Wang, L., Hu, W., Tan, T.: Recent development in human motion analysis. Pattern Recognition **36**(3), 585–601 (2003)
4. Khare, M. Kushwaha, A.K. S., Srivastava, R.K., Khare, A.: An approach towards wavelet transform based multiclass object classification. In: Proceeding of 6th International Conference on Contemporary Computing, pp. 365–368 (2013)
5. Sialat, M., Khlifat, N., Bremond, F., Hamrouni, K.: People detection in complex scene using a cascade of boosted classifiers based on Haar-like Features. In: Proceeding of IEEE International Symposium on Intelligent Vehicles, pp. 83–87 (2009)
6. Viola, P., Jones, M.: Rapid object detection using a boosted cascade of simple features. In: Proceeding of IEEE International Conference on Computer Vision and Pattern Recognition (CVPR), vol. 1, pp. 83–87 (2001)
7. Dalal, N., Triggs, B.: Histograms of oriented gradients for human detection. In: Proceeding of IEEE International Conference on Computer Vision and Pattern Recognition (CVPR), pp. 886–893 (2005)
8. Cao, X., Wu, C., Yan, P., Li, X.: Linear SVM Classification Using Boosting HoG Features for Vehicle Detection in Low-Altitude Airborne Videos. In: Proceeding of IEEE International Conference on Image Processing (ICIP), pp. 2421–2424 (2011)
9. Lu, H., Zheng, Z.: Two novel real-time local visual features for omnidirectional vision. Pattern Recognition **43**(12), 3938–3949 (2010)
10. Lowe, D.: Object recognition from local scale invariant features. In: Proceeding of 7th IEEE International Conference on Computer Vision (ICCV), pp. 1150–1157 (1999)
11. Yu, G., Slotine, J.J.: Fast Wavelet-Based Visual Classification. In: Proceeding of IEEE International Conference on Pattern Recognition (ICPR), pp. 1–5 (2008)

12. Khare, A., Tiwary, U.S.: Symmetric Daubechies complex wavelet transform and its application to denoising and deblurring. WSEAS Transactions on Signal Processing 2(5), 738–745 (2006)
13. Khare, M., Srivastava, R.K., Khare, A.: Single Change Detection based Moving Object Segmentation by using Daubechies Complex Wavelet Transform. Accepted in IET Image Processing (2013). doi:10.1049/iet-ipr.2012.0428
14. Renno, J.P., Makris, D., Jones, G.A.: Object classification in visual surveillance using Adaboost. In: Proceeding of IEEE International Conference on Computer Vision and Pattern Recognition (CVPR), pp. 1–8 (2007)
15. Chen, L., Feris, R., Zhai, Y., Brown, L., Hampapur, A.: An integrated system for moving object classification in surveillance videos. In: Proceeding of IEEE International Conference on Advanced Video and Signal Based Surveillance, pp. 52–59 (2008)
16. Castleman, K.R.: Digital Image Processing. Prentice Hall, Englewood Cliffs, NJ, USA (1996)
17. Selesnick, I.W., Baraniuk, R.G., Kingsbury, N.G.: The Dual-Tree Complex Wavelet Transform. IEEE Signal Processing Magazine 22(6), 123–151 (2005)
18. Kingsbury, N. G.: The Dual-Tree Complex Wavelet Transform - A New Technique for Shift Invariance and Directional Filters. In: Proceeding 8th IEEE DSP Workshop, Bryce Canyon (1998)
19. Teague, M.: Image analysis via the general theory of moments. Journal of Optical Society of America 70(8), 920–930 (1980)
20. Hwang, S.K., Kim, W.Y.: A Novel approach to the fast computation of Zernike moments. Pattern Recognition 39(11), 2065–2076 (2006)
21. Celebi, E. M., Aslandogan, Y. A.: A comparative study of three moment based shape descriptors. In: Proceeding of International Conference on Information Technology: Coding and Computing (ITCC 2005), vol. I, pp. 788–793 (2005)
22. Chong, C.W., Raveendran, P., Mukundan, R.: Translation invariance of Zernike moments. Pattern Recognition 36(8), 1765–1773 (2003)
23. Bin, Y., Xiong, P.J.: Invariance analysis of improved Zernike moments. Journal of Optics A: Pure and Applied Optics. 4(6), 606–614 (2002)
24. Papakostas, G.A., Boutalis, Y.S., Karras, D.A., Mertzios, B.G.: A new class of Zernike moments for computer vision applications. Information Sciences. 177(13), 2802–2819 (2007)
25. Farzem, M., Shirani, S.: A robust multimedia watermarking technique using Zernike transform. In: Proceeding of Fourth IEEE Workshop on Multimedia Signal Processing, pp. 529–534 (2001)
26. Zhenjiang, M.: Zernike moment based image shape analysis and its application. Pattern Recognition Letters 21(2), 169–177 (2000)
27. Khare, M., Srivastava, R.K., Khare, A.: Moving Object Segmentation in Daubechies Complex Wavelet Domain. Accepted in Signal Image and Video Processing (2013). doi:10.1007/s11760-013-0496-5
28. Noble, W.S.: What is Support Vector Machine. Nature Biotechnology 24(12), 1565–1567 (2006)
29. INRIA Person Dataset. http://pascal.inrialpes.fr/data/human (last accessed September 21, 2014)

Erratum to: On the Performance of Triangulation-Based Multiple Shooting Method for 2D Geometric Shortest Path Problems

Phan Thanh An[1,2], Nguyen Ngoc Hai[3], and Tran Van Hoai[4], and Le Hong Trang[1,5(✉)]

[1] Instituto Superior Técnico, CEMAT, Av. Rovisco Pais,
1049-001 Lisboa, Portugal
lhtrang@math.ist.utl.pt
[2] Institute of Mathematics, Vietnam Academy of Science and Technology (VAST),
18 Hoang Quoc Viet Road, Hanoi 10307, Vietnam
[3] Department of Mathematics, International University,
Vietnam National University, Thu Duc, Ho Chi Minh City, Vietnam
[4] Faculty of Computer Science and Engineering, HCMC University of Technology,
268 Ly Thuong Kiet Street, Ho Chi Minh City, Vietnam
[5] Faculty of Information Technology, Vinh University,
182 Le Duan, Vinh, Vietnam

Erratum to:

Chapter "On the Performance of Triangulation-Based Multiple Shooting Method for 2D Geometric Shortest Path Problems" in:

A. Hameurlain et al. (Eds.), *Transactions on Large-Scale Data- and Knowledge-Centered Systems XVI,*
DOI 10.1007/978-3-662-45947-8_4

The affiliation of "Phan Thanh An" was incorrect. The correct affiliation is "Institute of Mathematics, Vietnam Academy of Science and Technology (VAST), 18 Hoang Quoc Viet Road, Hanoi 10307, Vietnam."

The online version of the original chapter can be found under
DOI 10.1007/978-3-662-45947-8_4

A. Hameurlain et al. (Eds.): TLDKS XVI, LNCS 8960, p. E1, 2015.
DOI: 10.1007/978-3-662-45947-8_8

Author Index

An, Phan Thanh   45

Binh, Nguyen Thanh   87

Dang, Tran Khanh   1, 57
Dang, Tran Tri   1

Hai, Nguyen Ngoc   45
Hoai, Tran Van   45

Kawamura, Takahiro   15
Khare, Manish   87
Kosorus, Hilda   29
Küng, Josef   29

Le, Thu Thi Bao   57

Nguyen, Thi Ai Thao   57

Ohsuga, Akihiko   15

Quang-Hung, Nguyen   71

Son, Nguyen Thanh   71
Srivastava, Rajneesh Kumar   87

Thoai, Nam   71
Trang, Le Hong   45
Truong, Quynh Chi   57

Printed in the United States
By Bookmasters